Melting and Melt Movement
in the Earth

Melting and Melt Movement in the Earth

Proceedings of a Royal Society Discussion Meeting held on
3 and 4 March 1992

Organized and edited by

K. G. Cox, D. P. McKenzie, and R. S. White

THE ROYAL SOCIETY
OXFORD UNIVERSITY PRESS
1993

Oxford University Press, Walton Street, Oxford OX2 6DP

Oxford   New York   Toronto
Delhi   Bombay   Calcutta   Madras   Karachi
Kuala Lumpur   Singapore   Hong Kong   Tokyo
Nairobi   Dar es Salaam   Cape Town
Melbourne   Auckland   Madrid

and associated companies in
Berlin   Ibadan

Oxford is a trade mark of Oxford University Press

Published in the United States
by Oxford University Press Inc., New York

First published in Philosophical Transactions of the Royal Society of London
Series A, Volume 342, pp. 1–191.

A catalogue record for this book is available from the British Library

Library of Congress Cataloging in Publication Data

(Data applied for)

ISBN 0–19–854078–7

Typeset and printed in Great Britain by the University Press, Cambridge

Preface

For many years petrologists have believed that basaltic magmas are produced by partly melting the mantle with a composition similar to that of spinel and garnet perodotite nodules brought up as xenoliths in alkali basalts. The arguments were primarily chemical: laboratory experiments at high pressure showed that melts of basaltic composition can be produced by melting 5–30 per cent of such nodules. A few earth scientists, especially geophysicists, disagreed, and argued for many years that basalts are generated by complete melting of an eclogite with the same composition as basalt. Their view has, however, not been generally accepted, partly because it cannot account for the great range in the composition of igneous rocks. Though there has been widespread agreement among petrologists on the nature of the source region of basalts for at least 20 years, until 10 years ago there was little understanding of the fluid dynamical processes that allow the basaltic melt to separate from its residue. Fluid dynamics had been extensively used in petrology to understand the processes in basaltic magma chambers, whose evolution is strongly affected by compositional, and associated density, variations in the melt caused by the preferential removal of MgO. Such work depends on an understanding of the fluid dynamics of double diffusion, since both heat and compositional variations obey scalar diffusion equations.

Double diffusion is very important in oceanography, where the dynamics is influenced by both temperature and salinity variations. Many of the people involved in the oceanographic work were quick to see its relevance to magma chambers. In contrast, the fluid mechanics of melt separation had not been developed elsewhere. Several groups became interested in this problem in the 1970s and 1980s, and all used the ideas of mixture theory to derive various formulations of the governing equations. In mixture theory the melt and the matrix are both separately assumed to fill all space, and to obey the equations governing the behaviour of compressible viscous fluids. Their interaction is described by terms in the equations that allows mass, momentum, energy, and compositional variations to be transferred from one fluid to the other. The constraint that these terms should be independent of the frame of reference in which the equations are written (e.g. invariant to a gallilean transformation), and that the momentum equation should reduce to D'Arcy's law in the absence of flow in the matrix, are sufficient conditions to derive the governing fluid dynamical equations.

The same equations have been obtained by other arguments that make assumptions about the geometry of the channels in which the melt flows. Though the equations are complicated and nonlinear, they are not as mathematically intractable as they appear. An early result of great importance is that they allow the propagation of solitary waves (see Spiegelman, this volume). Indeed it is now widely believed that three dimensional solitary waves are the only stable three dimensional solutions to the equations, at least in the absence of a larger scale matrix flow. The existence of new equations that have solitary wave solutions has attracted considerable interest among fluid dynamicists, and is becoming a small specialized branch of the theory of two phase flow. But it is important to remember that the two phase flow equations derived using mixture theory do not describe the whole process of the separation of the melt from the residue. They are only valid if both the melt and matrix form interconnected networks in three dimensions. This condition is clearly not satisfied by most magmas when they are erupted, because interconnection of the matrix ceases when the melt fraction exceeds about 25 per cent. Most magmas contain a much smaller fraction of source material, but where the disaggregation of the matrix occurs is not known. Furthermore, it is hard to see how to study the process mathematically, because the governing equations change when disaggregation occurs. Little progress has yet been made in this important area.

In contrast, the processes that govern the interconnection of the melt are relatively well understood (see Harte, this volume), and are controlled by the surface energy between matrix crystals and the melt. Most primary magmas are now known to form interconnected networks at all melt fractions, no matter how small these may be. It is then straightforward to solve the two phase flow equations to discover how large the melt fraction needs to be before it can move at a geologically significant rate (~ 10 mm a^{-1}) with respect to the matrix, and so separate as a magma. For oceanic ridge basalts the result is ~ 1 per cent, but hot volatile-rich magmas such as lamproites and kimberlites can probably separate from their source regions at melt fractions that are as small as 10^{-2} per cent. When these results were first obtained they were consistent with existing geochemical arguments which required very small melt fractions to separate from their residues. They were not, however, consistent with the widely held views of experimental petrologists, who thought that the melt fractions had to exceed 5–10 per cent before movement could occur. As far as I can discover this view was intuitive, and was not based on an understanding of the melt separation process. It is now rarely heard. Though most petrologists now accept that very small melt fractions can separate from the residue, one aspect of the process still often causes confusion. If small melt fractions can separate, they must *always* do so in all environments in which melting occurs. This result is obvious: when a small amount of melting has occurred there is no way in which the source region can know whether it will undergo more melting. Therefore, though the demonstration that MORB is generated by fractional melting is important, it is scarcely surprising.

There is as yet almost no geophysical information about regions of the Earth in which melt is being generated, whereas there is a wealth of chemical information about the composition of the melts. For a number of reasons the chemical data is not easy to interpret. The composition of the magma at the surface is some sort of average of that of the melt being generated, but, without a fluid dynamical model for the disaggregation, it is impossible to know how to calculate the average. Furthermore, the melt may not everywhere be in chemical equilibrium with the residue. Given these problems, most attempts to model melt composition have calculated the composition of the pooled melt from the whole melting region, and have compared this with the observed mean magma composition. In the case of the trace elements it is straightforward to calculate the melt composition generated by fractionation melting, using either simple analytic expressions or numerical integration along the melting path. In contrast similar expressions are not available for major elements. Takahashi *et al.* (this volume) describe the results from a new set of experiments in which the melt fraction and the composition are measured. Though their new techniques represent a major improvement on earlier experiments, their results from batch melting still cannot be used to calculate the major element composition of a fractional melt. For this reason trace element concentrations are likely to remain the most straightforward way of comparing fluid dynamical models with geological observations. Given these difficulties it is obviously of great importance to use geophysical methods to observe regions that are melting. The two techniques that show most promise are the measurement of electrical conductivity and of seismic velocity. Though the electrical conductivity is very strongly affected by the presence of interconnected melt, it is not straightforward to calculate the three dimensional variation of conductivity from surface measurements. In contrast, it is easier to invert seismic travel times to obtain a three dimensional velocity model, but harder to relate velocity variations to melting processes.

Preface

Though we are beginning to understand the intimate relationship between geodynamical processes and melt generation, a major remaining problem is that igneous petrologists and geophysicists have rarely worked closely together, and the skills of both groups are required to understand the processes involved. However, as the papers in this volume illustrate, these two groups at present think quite differently, and have difficulty understanding each other. Three of us therefore thought that we would organize this Discussion Meeting to bring together some of those involved in these interconnected lines of research.

Oxford and Cambridge   K. G. C.
March 1993   D. P. McK.
                            R. S. W.

Contents

Contributors

Dr B. Harte *Grant Institute of Geology, Edinburgh*

Dr M. Spiegelman *Lamont–Doherty Geological Observatory of Columbia University, Palisades, New York*

Professor D. L. Kohlstedt *Department of Geology and Geophysics, University of Minnesota, Minneapolis*

Dr G. Ceuleneer *GRGS–CNRS, Toulouse*

Professor R. K. O'Nions *Department of Earth Sciences, University of Cambridge*

Professor J. B. Gill *Department of Geology, Universite Blaise Pascal, Clairmont-Ferrand*

Mr M. Kurtz *Department of Chemistry, Woods Hole Oceanographic Institution, Massachusetts*

Professor E. Takahasi *Earth and Planetary Sciences, Tokyo Institute of Technology*

Professor F. A. Frey *Massachusetts Institute of Technology, Cambridge, Massachusetts*

Professor R. S. White *Department of Earth Sciences, University of Cambridge*

Professor K. G. Cox *Department of Earth Sciences, University of Oxford*

Dr A. Hasegawa *Observation Centre for Prediction of Earthquakes and Volcanic Eruptions, Faculty of Science, Tohoku University, Sendai*

Professor C. J. Hawkesworth *Department of Earth Sciences, The Open University, Milton Keynes*

1. Introduction

This paper aims to consider the behaviour of basic-ultrabasic silicate melts in the mantle, and how they are related to the formation of metasomatic rocks and alkalic igneous rocks in the mantle lithosphere. The association in the mantle of modally metasomatized rocks with dykes and veins indicative of melt injection is emphasized. Particular attention is given to the way in which the crystallization of metasomatic minerals may be linked with the chemical evolution of melts injecting and percolating through peridotite, so that continuous varieties of both metasomatized rocks and melts responsible for one another may be formed. Evidence for these continua of variations of metasomatic rocks and associated melts is given by both petrographic and trace element data. The suggestions put forward are dependent upon the ability of basic and ultrabasic melts to move through peridotite matrices even when present in small volumes (less than 1–5 %). The evidence relating to this, from theoretical and experimental standpoints as well as information from natural rocks (especially cumulates), is summarized in §2.

2. Melt geometry and melt mobility: theory, experiment and natural rock textures

The principles governing the shape of mineral grains and pores filled with fluids (including melts) in an aggregate of phases are well understood (Smith 1948; Voll 1960; Beere 1975; Kingery *et al.* 1976), and their importance in controlling melt and other fluid distribution in rocks has been the focus of much recent attention (Bulau *et al.* 1979; McKenzie 1984; Watson & Brennan 1987; Cheadle 1989; and papers quoted in table 1).

In simple terms the important question is whether small amounts of melt will tend to disperse along grain edges (as in figure 1*a*), or form isolated pockets (as in figure 1*b*). Under conditions of textural equilibrium, the melt distribution is controlled by the relative surface tension (or interfacial energy) of the interfaces of melt against solid by comparison with the solid–solid interfaces. These relative interfacial energies determine the geometrical relations of the melt pores with the solid grains, and are most simply expressed in terms of the dihedral angle formed by a melt pool in contact with two solid grains.

If the dihedral angle for melt between minerals is less than 60°, the melt disperses to form curved prismatic channels along the triple junctions between mineral grains (figure 1*a*). This means that the melt forms a continuously connected phase throughout the rock even if the melt is present in only vanishingly small amounts. Where the melt dihedral angle is more than 60°, small melt volumes will form isolated pores at grain corners and edges as in figure 1*b*. In this situation the establishment of continuous connection (connectivity) for the whole melt fraction depends on increasing the proportion of melt in the system until the pores of figure 1*b* meet one another. Figure 1*c* shows the dependence of melt connectivity on melt volume fraction and dihedral angle; with greater amounts of melt being necessary to establish connectivity as the dihedral angle increases. Where several solid phases are present, differences in dihedral angle may result in connectivity of melt being established between certain phases but not others (Toramaru & Fujii 1986).

Table 1 (after Cheadle 1989) lists experimentally determined dihedral angles for melts and solids of current interest. Most of the experiments have been done with

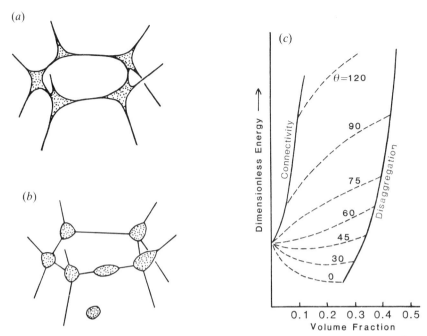

Figure 1. Textural and distribution relationships for a fluid phase. (*a*), (*b*) Melt pore shapes (dotted) in a three-dimensional aggregate of mineral grains for dihedral angles respectively below and above 60°. In (*a*) the melt extends along the grain edges between mineral grains. In (*b*) it forms isolated pools on grain edges or faces. (*c*) The volume fractions of melt necessary for melt 'connectivity' (continuous porosity) and 'disaggregation' (solid grains no longer in continuous contact) with various dihedral angles (after Beere 1975). θ is the dihedral angle.

basaltic melt and some contain H_2O and/or CO_2. Although there is some variation, it is clear, at least for olivine and probably orthopyroxene matrices, which dominate mantle rocks, that the dihedral angles for basaltic, komatiitic and carbonatitic melts are less than 60° and probably mainly in the range 25–40°. These data indicate that even very small melt fractions in the mantle should show connectivity and readily disperse through the rocks.

The formation of a connected three-dimensional melt network facilitates the relative upward movement of the melt with respect to its solid matrix under gravity. The length and timescales for gravity-driven compaction (McKenzie 1984, 1985) depend principally upon the material properties of the system: matrix and melt viscosities, melt density, and the permeability (which depends on dihedral angle; see Cheadle 1989). When the partly molten region is large in comparison to the compaction length, as it is in partly molten mantle, the compaction timescale is largely a function of melt viscosity. Basaltic melts (viscosity greater than 1 Pa s) will only separate from their matrix on reasonable geological timescales, if the melt fraction is greater than 1%; volatile-rich melts such as carbonatites with viscosities of *ca.* 10^{-1} Pa s will separate with melt volumes as low as 0.1% (O'Nions & McKenzie 1988; Hunter & McKenzie 1989).

The widespread mobility of mantle melt fractions of around 1% or less is clearly significant, because of its effect in concentrating and redistributing the incompatible elements which preferentially dissolve in the melt (McKenzie 1984, 1989). In basalt petrogenesis it provides a mechanism for small volumes of melt to extract

Table 1. *Summary of experimentally determined solid–solid–melt dihedral angles*

(ol, olivine; opx, orthopyroxene; cpx, clinopyroxene; ga, garnet; am, amphibole; pl, plagioclase; bi, biotite; af, alkali feldspar; qz, quartz; bas and, basaltic andesite; alk bas, alkali basalt.)

solid phase contact	melt composition	pressure (GPa)	temp. °C	dihedral angle	ref.[a]
ol–ol	basalt	1.0	1240	47	1
ol–ol	basalt	1.0	1230	22–35	2
ol–ol	basalt	1.0	1316	20–27	2
ol–ol	basalt	1.0	1360	21–26	2
ol–ol	basalt	1.5	1316	23–27	2
ol–ol	basalt	2.0	1316	29–38	2
ol–ol	basalt	2.0	1350	25–39	3
ol–ol	basalt	1.0	1300	49	4
ol–ol	basalt	1.5	1250	41	5
ol–ol	komatiite	0.0001	1400	32	6
ol–ol	alk bas	0.8	1250	46	6
ol–ol	basalt	0.3	1050	45	7
ol–ol	basalt	0.3	1255	43	7
ol–ol	komatiite	0.0001	1450	29–33	8
ol–ol	carbonate	3.0	1290–950	28	9
ol–opx	basalt	2.0	1350	24–40	3
ol–opx	basalt	1.5	1250	59	5
ol–opx	carbonate	3.0	1290–950	< 60	9
ol–ga	carbonate	3.0	1290–950	< 60	9
ol–am	bas and	0.8	1050	38	10
opx–opx	basalt	1.5	1250	52	5
opx–opx	basalt	1.0	1250	20–23	3
opx–opx	basalt	1.5	1350	29–34	3
opx–opx	basalt	2.0	1350	36–41	3
opx–opx	basalt	2.5	1350	33–40	3
am–am	bas and	0.8	1050	33	10
am–pl	bas and	0.8	1050	54	10
pl–pl	bas and	0.8	1050	60	10
bi–bi	silicic	1.2	1050	30	11
af–af	felsic	1.0	1000	44	12
af–qa	felsic	1.0	1000	49	12
qz–qz	felsic	1.0	1000	59	12

[a] References: (1) Waff & Bulau 1979. (2) Bulau 1982. (3) von Bargen & Waff 1988. (4) Toramaru & Fujii 1965. (5) Fujii *et al.* 1986. (6) Jurewicz & Jurewicz 1986. (7) Riley & Kohlstedt 1991. (8) Walker *et al.* 1988. (9) Hunter & McKenzie 1989. (10) Vincenzi *et al.* 1988. (11) Laporte 1988. (12) Jurewicz & Watson 1985. Based on a compilation by Cheadle (1989).

incompatible elements from a large volume of mantle. It also has application to many aspects of mantle metasomatism (Bailey 1982; Harte 1983; Menzies 1983) because it means that small volume melts can establish connectivity and move through rocks in a manner similar to that often envisaged for hydrous or other metasomatizing fluids. Even when relatively static, the dispersed and connected melt will provide a relatively fast transport medium for movement of material by diffusion compared to lattice diffusion through mineral matrices. Thus many mantle metasomatic and enrichment phenomena, which have been previously attributed to volatile-like fluids may potentially be products of infiltration by small-volume melts (Harte 1987). Indeed, where melts are present, metasomatism should be commonplace, since surface tension and gravity forces will cause melts to move along

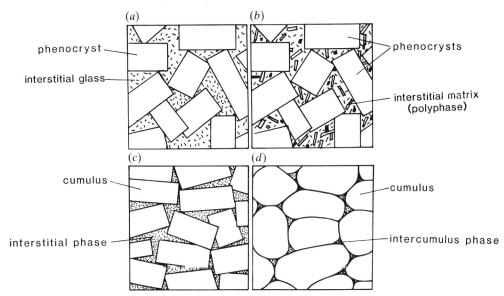

(a)

phenocryst

interstitial glass

(b)

phenocrysts

interstitial matrix (polyphase)

(c)

cumulus

interstitial phase

(d)

cumulus

intercumulus phase

Figure 2. Schematic illustrations of textures in igneous rocks. (a), (b) Euhedral shapes of phenocrysts within a matrix whose overall shape mimics that of the pre-existing melt, which has been quenched to a glass in (a), and consists of a rapidly cooled aggregate of fine crystals in (b). (c) Typical cumulate texture with a cumulus mineral (whose euhedral-subhedral shape reflects growth within melt) and a substantial volume of intercumulus space occupied by only one other mineral. (d) A stage in the progression from (c) towards a monomineralic cumulate, with the cumulus mineral grains having substantial impingement upon one another and now showing granuloblastic (polygonal-granoblastic) shapes whilst the intercumulus volume is severely reduced, but still consists of only one phase.

grain edges in very small volumes and affect compositional alteration, without necessarily causing any obvious petrographic change.

In considering the theoretical and experimental evidence summarized above, a legitimate question has to concern the extent to which natural situations involving melt achieve local textural equilibrium and hence the predicted pore-system microgeometries. Consideration of this also raises important questions of how former melt is recognised once crystallization occurs, and to what extent localized crystallization products of the melt reflect its bulk composition.

Typical textures in igneous rocks, which clearly preserve evidence of former melt geometry, are often not symptomatic of textural equilibrium. The euhedral shapes of crystals preserved in a quenched glass or rapidly chilled fine-grained crystalline matrix (figures 2a, b), are a product of crystal growth mechanisms in melt rather than textural equilibration. In many basalts and dolerites such intersertal and interstitial textures are common, and where the matrix is glass or comprises many phases it probably approximates to the host melt composition. In other cases, as in many gabbros and cumulates, a single mineral forms the interstitial matrix phase (figure 2c). This single interstitial or intercumulus phase is typically one of the common silicate minerals and it is evident that it can in no way provide a reasonable estimate of the consanguineous melt composition. Yet the texture (figure 2c) is essentially similar to that where the interstitial matrix is glass or a polyphase aggregate (figure 2a, b), and it is evident in all cases that the interstitial matrix has taken on the melt shape and can be loosely described as a *melt pseudomorph*. The melt shape assumed by the interstitial/intercumulus phase is not a pseudomorph of the melt at one instant of time, but is a relict of melt shape formed over an interval of time during which cumulus and intercumulus phases were growing.

The texture highlighted in figure 2c, where a single interstitial phase pseudomorphs melt geometry is common over a wide variety of situations (Hunter 1987; Harte et al. 1991 a). Clearly during the crystallization of the single interstitial/intercumulus phase, its chemical components must have become concentrated in the particular intercumulus volumes concerned, while other melt components, destined for other phases, migrated elsewhere. Such migration must have remained efficient even with very small melt fractions and will clearly have been facilitated by the connectivity of the melt along grain edges predicted by the theoretical and experimental data discussed above. Particular evidence of such inter-connected porosity at small melt volumes is seen in rock textures where the interstitial material forming several intercumulus pools is a single crystal (oikocryst) even though such pools appear disconnected in two dimensions. The transport of melt components through the intercumulus volume will also be considerably aided by relative movement (two-phase flow) of melt and crystalline material, which may be expected to occur by gravity-driven compaction in situations such as typical layered cumulate bodies and the mantle (McKenzie 1984, 1989; Hunter 1987).

In seeking analogies with mantle conditions, basic-ultrabasic layered cumulate bodies provide excellent examples of situations where we may observe minerals and melts kept at high temperatures for relatively long periods of time, with melt percolating through a crystalline matrix. Cumulates also give evidence of a variety of textures formed under differing conditions of porosity (melt fraction), including situations where the amount of melt is very small. Monomineralic cumulates represent an extreme example of rocks where efficient transport of melt components, even at very small melt fractions, has allowed only one phase to form. In cumulates, progression towards very low abundances of intercumulus minerals, commonly goes hand-in-hand with progression towards textural equilibrium, as illustrated by the textures of figure 2c and d and the type monomineralic accumulates of Wager et al. (1960). The changes in grain shape associated with this progression and the compaction accompanying it may be affected by a number of mechanisms (e.g. solution-reprecipitation, diffusion, dislocation creep), but the result, under the usual conditions of little crystalline strain, is always to produce a lower energy grain geometry. This progression towards textural equilibration will, therefore, ensure that the melt maintains an inter-connected porosity as the melt fraction becomes small, and thereby will allow migration of melt components even at very small melt fractions.

The percolation of intercumulus melt by compaction in large basic and ultrabasic intrusions is likely to be applicable to the movement of small fractions (1–5%) of melt through partly molten regions within the mantle. It is notable that just as textural evidence of thorough-going textural equilibrium throughout large rock volumes is limited in layered cumulate bodies, so it is in the mantle. Many of the common coarse mantle xenoliths show relatively poorly developed granuloblastic textures and therefore have only a moderate degree of textural equilibrium between mineral grains (Harte 1977). Thus for melt connectivity and mobility to have been established in such rocks, there can only have been localized textural equilibrium along grain edges as seen in some layered cumulates. Specific evidence supporting this is seen in the wallrocks to many mantle dyke-like bodies (Wilshire & Shervais

Table 2. *Modal metasomatic associations*

erupting host	principal metasomatic mineral in peridotites	associated dykes and veins at depth
(1) highly alkaline volcanics in continental rifts	*clinopyroxene, biotite, amphibole* titanomagnetite, sphene, apatite (calcite, feldspar)	*clinopyroxenites* (with apatite and titanomagnetite)
(2) alkali basalt– basanite–nephelinite series	*kaersutite/pargasite, biotite,* apatite, ilmenite, clinopyroxene	Al–augite wehrlite–pyroxenite dykes, amphibole–mica (lherzite) veins
(3) kimberlite	*biotite, K-richterite, clinopyroxene,* K–Ba–titanates, ilmenite, rutile, apatite, zircon. (*Marid type*)	MARID suite dykes and biotite–richterite veins
(4) kimberlite	*ilmenite, rutile, biotite, sulphides,* clinopyroxene, garnet (*Matsoku type*)	opx- and cpx-rich dykes and veins, (with garnet and IRPS minerals)
(5) kimberlite	*edenite, biotite,* clinopyroxene (*Jagersfontein type*)	rare zones rich in cpx and/or amphibole and/or biotite

1975; Irving 1980; Harte *et al.* 1987), where there is chemical compositional evidence of melt infiltration, despite a lack of textural equilibrium amongst the crystalline phases, or any petrographic evidence of the passage of the melt. Recently, Matthews *et al.* (1992) have provided further evidence from the occurrence of healed fractures in garnets in such wallrocks, that melt must have infiltrated these rocks despite the fact that the petrographic textures show only a moderate approach to textural equilibrium between mineral grains.

3. Mantle dykes, veins and metasomatic rocks: xenolith petrography

(a) *Associations of dyke, vein and modal metasomatic rocks*

The occurrence of metasomatism in mantle rocks is now widely accepted and there is considerable evidence to show that such phenomena are in situ mantle processes, and not merely the result of alteration during and/or following eruption (Harte 1983, 1987; Menzies 1983; Dawson, 1984). Metasomatism may take place with or without changes in the petrography of the rocks, and where such changes occur they do so both by introduction of new phases such as biotite, amphibole, ilmenite, rutile and apatite, and by changes in the proportions and compositions of those typical peridotite–pyroxenite minerals originally present (i.e. olivine, pyroxenes and garnet). Such modal metasomatism may be classified according to the nature of the minerals occurring and their associations, and the principal types of metasomatism found in mantle xenoliths are summarised in table 2 (after Harte & Hawkesworth 1989).

As indicated in table 2, mantle rocks exhibiting modal metasomatic mineralogies are commonly found alongside others bearing direct evidence of mantle melt/fluid interaction in the form of intrusive dykes and veins, and it is important to consider these phenomena in conjunction (Gurney & Harte 1980; Harte 1983; Wilshire 1987). Thus we shall briefly review the variety of dyke and vein phenomena and their relation to modal metasomatic rocks.

The most widely reported phenomena providing evidence of intrusive sheet and vein-like bodies traversing mantle peridotite are from xenoliths erupted with alkali basalts. They show a wide range of characteristics with gradations towards the features seen in xenoliths from highly alkaline continental rifts (Menzies 1983; Lloyd 1987; Witt & Seck 1987). The fragmentary nature of the specimens retrieved from alkali basalts means that the overall relationships have to be pieced together. However, studies of composite xenoliths, and comparisons with relationships seen in alpine peridotite massifs, have made it possible to construct clear pictures of the dyke/vein relations to the country-rock peridotites, determine orders of intrusion, and suggest mechanisms of emplacement involving hydraulic fracturing (Wilshire & Pike 1975; Irving 1980; Wilshire *et al.* 1980; Nicolas & Jackson 1982; Wilshire & Shervais 1985). The dykes and veins in basalt-derived xenoliths vary widely in modal mineralogy, but are usually dominated by either clinopyroxene (typically Al–Ti–augite) or amphibole (kaersutite–pargasite) and as such may be relatively anhydrous or hydrous. The more abundant clinopyroxene-rich varieties form the Al augite wehrlite pyroxenite (or Type II) xenoliths (see reviews of Menzies 1983; Harte & Hawkesworth 1989), and there is a continuum of compositions between these and amphibole or amphibole–mica (lherzite) dominated veins (Wilshire & Trask 1971; Wilshire *et al.* 1980; Wilshire 1987), some of which carry substantial apatite (Wass *et al.* 1980). The full range of commonly occurring minerals is: clinopyroxene, orthopyroxene, olivine, amphibole, biotite, spinel, apatite, ilmenite and magnetite. Plagioclase may occur in lower pressure varieties, while there are also some garnet pyroxenite xenoliths which seem to be related (Frey 1980; Griffin *et al.* 1984). There are close counterparts to minerals of the Type II dykes/veins in the so-called megacryst minerals, which occur as xenocrysts in alkali basalt, and have been considered to represent phenocrysts from deep-seated magmas.

Dykes and veins are not so common in the kimberlite-sampled mantle material, but those that have been described show similar structural and textural features to the basalt-derived material, despite differences in mineralogy. Thus analogues of the basalt-derived lherzites are seen in veins rich in amphibole and mica, which have been extensively described traversing peridotite in xenoliths from the Kimberley area (Jones *et al.* 1982; Dawson 1987; Erlank *et al.* 1987). In these veins the amphibole is a K-richterite, and there is an association with a diverse suite of MARID intrusive sheets/dykes in kimberlites (table 2) consisting of widely variable proportions of MARID minerals (mica, amphibole, rutile, ilmenite, diopside), together with apatite, zircon and unusual titanates (Dawson & Smith 1977; Dawson 1987; Haggery 1983; Waters 1987). Similarities to the more anhydrous Type II dykes and veins from basalts are shown by the Matsoku xenolith suite (Harte *et al.* 1987). This includes dykes/sheets which show widely varying proportions of the common peridotite minerals (olivine, orthopyroxene, clinopyroxene, garnet) as well as *IRPS* minerals (ilmenite, rutile, phlogopite and sulphides), which are also prominent in associated metasomatic rocks. Similar metasomatic phenomena, where ilmenite is prominent, have been described from other kimberlites (Wyatt & Lawless 1984), and there are compositional links also between the Matsoku suite and the common Cr-poor megacryst suite in kimberlite (Harte *et al.* 1991 *b*) and with the metasomatism affecting high-temperature deformed peridotites (Gurney & Harte 1980; Harte 1983). The Jagersfontein metasomatic suite, characterized by edenite and mica (table 2), shows the least evidence for dyke and vein structures (Field *et al.* 1989; Winterburn *et al.* 1990) though rare mica veins have been found (P. A. Winterburn, personal communication).

Examination of material from localities where both dyke/vein and modally

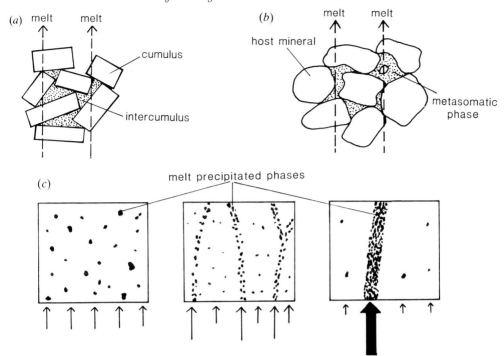

Figure 3. (*a*), (*b*) Comparison of the situations for normal cumulates and metasomatic infill cumulates. In both, a melt phase is percolating upwards, and in the former crystallizes the intercumulus phase which fills the original melt porosity between the gravity-settled cumulus crystals. In (*b*) the percolating melt deposits a metasomatic mineral phase in the melt porosity, which in this case lay within a grain network formed by host peridotite minerals. (*c*) The outlined boxes represent cross sections of peridotite within which minerals are being deposited from a melt flowing/percolating upwards; various degrees of channelizing of the melt flow are shown resulting in variations from pervasively metasomatized peridotite, through veined peridotite to dyke intruded peridotite. Arrows at bottom indicate relative variation in input of melt, which will vary over time as minerals are deposited by the melt.

metasomatized rocks are abundant, shows that these rock types appear to be very closely related. The evidence may be summarized as follows:

(*a*) The minerals introduced into modally metasomatized peridotites are commonly found in dykes and veins, where either they show similar chemical compositions or there are gradients in chemical compositions between peridotite and dyke/vein.

(*b*) A close spatial association may be visible in individual xenoliths (e.g. peridotites carrying veins are also often pervasively metasomatized with the widespread development of a vein mineral).

(*c*) Variations may be found from discrete dykes and veins to situations with more irregular concentrations of dyke/vein/metasomite minerals, to a completely dispersed distribution of dyke/vein/metasomatic phases (figure 3*c*).

(*d*) The dykes/veins from all the associations are extremely variable in modal composition, extending to virtually monomineralic rocks (e.g. clinopyroxenites, orthopyroxenites, amphibolitites, glimmerites), and these have their counterpart in wide modal variations in the metasomatized rocks (e.g. 'peridotites' containing abundant mica, or amphibole, or ilmenite).

(*e*) The host rock phases in metasomatized peridotites (e.g. olivine, pyroxenes, garnet) show changes in their chemical composition (e.g. Fe/Mg, Al/Cr) towards compositions found in the dykes/veins.

(b) *Textures and structures in dykes, veins and modal metasomatic rocks*

Both the dyke/vein and their associated modal metasomatic rocks show a wide variety of textures, which are reminiscent of the range of textures seen in cumulates, from ones clearly showing igneous textures to ones showing good solid–solid textural equilibrium. Thus in dykes, one may find euhedral/subhedral crystal shapes, interstitial and poikilitic/oikocrystic textures, and granuloblastic (polygonal granoblastic) textures (Wilshire & Shervais 1975; Frey & Prinz 1987; Harte *et al.* 1987; Harte & Matthews 1989). In the last type equant polygonal grains suggest a high degree of textural equilibrium, and are most common in monomineralic portions of the dykes, suggesting a parallel with accumulates. In some cases replacement textures suggest that multiple injections of melt/fluid have occurred along the dykes (Wilshire *et al.* 1980; Irving 1980).

In some cases in modal metasomatic rocks, there is also clear evidence of replacement of a pre-existing phase (usually olivine, orthopyroxene or clinopyroxene) by a metasomatic phase (usually amphibole or mica); (see Erlank *et al.* 1987, figs 6 to 8; Winterburn *et al.* 1990, fig. 2). Sometimes the metasomatic replacing phase has a dispersed poikilitic or oikocryst habit, which bears a close resemblance to the way the intercumulus phase in many cumulates shows crystallographic continuity across a series of interstitial spaces, and thereby provides evidence that the metasomatic phase crystallized from the melt/fluid. Occasionally, euhedral inclusions of olivine or orthopyroxene occur, within the metasomatic phases (Harte *et al.* 1987) and provide further evidence of crystallization from a melt. In other cases the metasomatic phases show a much closer approach to textural equilibrium characteristics, having smoothly curving grain boundaries and regular triple junctions. Such cases are interpreted as evidence that the metasomatic minerals reached textural equilibrium under near-solidus conditions as in the case of accumulates.

More particular evidence of the growth of metasomatic mineral aggregates from melt pools within peridotite is shown by ilmenite aggregates (see fig. 10, Harte *et al.* 1987) in the Matsoku peridotites (Harte & Matthews 1989). The ilmenites were found to have a broad and partly bimodal spectrum of dihedral angles against olivine and orthopyroxene: smaller angles (close to 60°) apparently matching those of melts, larger angles (near 100°) conforming to experimentally determined equilibrium angles for ilmenite-silicate textures (Matthews & Harte 1989). The Matsoku ilmenite aggregates are likened to the interstitial phase in many cumulates, originating by crystallization from melt pockets and retaining shape features which are partly those of the melts, while simultaneously moving towards textural equilibrium for the solid phases. The development of such features is considered to be dependent on the flow of melt through the rocks, which allows crystallization of only one phase while other melt components are carried away by the flowing melt (Harte & Matthews 1989).

In the dykes occurring with the various modal metasomatic associations it is quite common to find layering parallel to their walls (Irving 1980; Wilshire *et al.* 1980; Harte *et al.* 1975, 1987; Dawson & Smith 1977). Such layering has been attributed to magma flow and crystal plating on conduit walls, and although the overall features indicate more varied situations than crystal plating alone (Wilshire *et al.* 1980; Harte & Hunter 1986), it clearly indicates processes whereby crystals and melt become selectively separated.

(c) Crystallization processes in dykes, veins and modal metasomatic rocks

For all of the dyke/vein associations listed in table 2, an origin by silicate melt intrusion has been widely accepted for the larger dyke-like bodies, because of the involvement of a wide array of both anhydrous and hydrous minerals. However, for smaller vein-like bodies and for metasomatized rocks rich in hydrous phases, the nature and composition of the metasomatizing fluid is more debatable with many authors favouring hydrous volatile fluids (Menzies *et al.* 1987). We suggest that the continuum of wide variations in modal mineral proportions, both hydrous and anhydrous, in various dyke/vein/metasomite associations (table 2), provides no evidence for distinguishing ones which have crystallized from a hydrous fluid rather than a melt. Because of this and the fact that melts may yield both hydrous (e.g. amphibole-rich) and anhydrous (e.g. clinopyroxene-rich) precipitates, while non-molten fluids have limited element transport capabilities (Eggler 1987), we further suggest that the dyke/vein/metasomite rocks are in general melt-derived. Further evidence in support of the possibility that all dykes/veins/metasomites are melt-derived is given on the basis of the trace element data presented in the next section.

In considering the crystallization processes of the dyke/vein bodies, it is important to take into account the wide variations in mineral proportions found in closely related dykes/veins in all the associations described, coupled with the internal layering structures, and the similar range of textural features to those found in cumulates. All suggest that the dyke mineralogy is controlled by dynamic and kinetic factors of two-phase flow and crystallization. Under such circumstances, the crystal aggregates clearly do not represent bulk melt compositions, but are aggregates and residues deposited and left behind by the flowing melt. Thus the assemblages in the dykes and veins are analogous to layered intrusion cumulates in the sense that they represent crystal residues reflecting processes of crystal sorting, melt flow and percolation, and kinetic controls (e.g. nucleation and growth). At the same time they are certainly not simple products of crystal settling and for this reason, it is proposed to call them *dyke cumulates*. It is pertinent to note that their bulk trace element geochemistry is also similar to that of normal cumulates in reflecting the loss of incompatible components carried away with the melt (Frey & Prinz 1978; Irving 1980; Waters 1987).

Modal metasomatic rocks are envisaged as products of melt infiltration and injection, where, in comparison to the dyke and vein facies, the melt was more dispersed and percolating through a peridotite matrix (figure 3). In addition to any direct replacement reactions, the metasomatic phases are interpreted to have grown to fill the melt porosity (which might be less than 1.0% or up to about 30%) that existed when melt was flowing through the rocks, in similar manner to the way intercumulus material replaces space previously occupied by melt (figure 3). As the matrix in this instance is the host peridotite, the analogy to cumulates made here does not so much involve the cumulus crystals themselves as the intercumulus material, and for this reason we refer to the metasomatic rocks as *metasomatic infill cumulates* (figure 3). Note that the metasomatic phases infilling the melt porosity, may be both monomineralic and polymineralic, in a similar way to intercumulus material. Where percolation and transport processes combine with crystal nucleation processes to produce monomineralic metasomatic material, the effects upon bulk rock composition are much more striking, but the metasomatic material necessarily gives a poor guide to bulk melt composition.

In addition to the crystal deposition, infiltrating melts would be expected to react and equilibrate with the host peridotite, thereby causing an exchange of elements with the matrix crystalline phases according to chromatographic column and percolation principles (Hoffman 1973; McKenzie 1984; Navon & Stolper 1987; McKenzie & O'Nions 1991). The flowage of melt through the rocks, therefore, represents a complex process in total. On the one hand there is modification of the melt by interchange with the host rock according to a percolation exchange process. On the other hand the melt is changing as a result of precipitating and leaving behind the metasomatic phases, which is a type of crystal fractionation process. The combination of processes can thus be described as '*percolative fractional crystallization*'. The difference of this melt differentiation process to normal crystal fractionation is that the melt is maintaining at least partial equilibrium with the host rock phases it is percolating through. Maintenance of equilibrium with olivine, the most abundant and ubiquitous of host rock phases, is particularly to be expected and will mean the melt remains rich in compatible elements such as Ni and Co.

4. Use of trace element data on minerals from dykes and metasomatic rocks to determine mantle melt compositions

It is difficult to know the percentage of melt that may have been present at any one time in the metasomatized rocks, or the amount of melt that may have passed through the dykes/veins/metasomites and the extent to which equilibrium with host rock phases was maintained. Clearly the processes described above result in differentiation of the flowing and percolating melt, and leave the dyke, vein and metasomatic phases as residua. Thus bulk rock compositions for dykes/veins/metasomites will be no better guide to melt compositions than bulk cumulate rock compositions are to the melts responsible for them. To reconstruct melt compositions it is therefore imperative to use mineral compositions that coexisted with the melts and calculate the melt composition with the use of appropriate partition coefficients. This approach offers particular promise with trace elements.

The amount of mineral trace element data from xenoliths, as distinct from whole rock and major element data, that is available in the literature is quite small (see particularly Shimizu (1975) and Kramers *et al.* (1981) for kimberlite-derived peridotites and megacrysts; Kramers *et al.* (1983) for MARID related rocks; Wass *et al.* (1980), Irving & Frey (1984), Menzies *et al.* (1985) for basalt-derived xenoliths and megacrysts). Much of this data was compiled and reviewed by Menzies *et al.* (1987), who also calculated some melt compositions from them. Their compilations showed considerable similarities, with generally gradual variations in concentrations, between trace element distribution patterns for mineral phases derived from dykes, veins and metasomatic rocks as well as from megacrysts.

We have augmented these data with new analyses of clinopyroxene, amphibole, garnet and mica from the Matsoku, Jagersfontein and Eifel xenolith suites. The new data have been gathered by using the Cameca ims 4f ion microprobe at Edinburgh University. This technique consists of routine *in situ* SIMS analysis of various REE, LILE and HFSE derived from a 20–30 μm spot on the target mineral by bombardment with an 8–10 nA oxygen ion beam, and using a 100 eV energy offset to minimize isobaric spectral interferences (for analytical details see Matthews *et al.* (1992)). Elemental abundances are normalized against analyses of NBS610 glass, and validated by analyses of mineral standards. Detection limits for the REE range from 10 to 100 p.p.b. The full data-set will be published elsewhere.

Melt compositions have been calculated, from this new data and those already in the literature for clinopyroxenes and amphiboles, for the spectrum of dyke, vein and metasomatic rock associations given in table 2. The partition coefficients used are those adopted by McKenzie & O'Nions (1991), coupled with use of the same values for Y and Zr as for Ho and Hf respectively; together with the addition of a value of 0.5 for Ba between amphibole and melt based on Irving & Frey (1984), and a guessed one of 0.05 for Ba distribution between clinopyroxene and melt. Representative calculated melt compositions are shown in figure 4. These are normalized to chondritic abundances taken from Sun & McDonough (1989).

Figure 4*a* shows calculated melts coexisting with clinopyroxenes from dykes/veins/metasomatic rocks and a megacryst collected from kimberlite sources. Although LREE vary in concentration by an order of magnitude, the patterns show considerable regularity, and reveal a fanning array, with progressively increasing LREE-MREE away from the melt composition coexisting with the megacryst (MEGA 1600E). The megacryst melt composition has been shown to be consistent with that expected for a primitive asthenospheric OIB-like melt (Harte 1983; Jones 1987). Thus the progressive enrichment in LREE and MREE seen in the metasomatic melt compositions in figure 4*a* may be generated by progressive extraction of metasomatic phases (especially clinopyroxene, garnet and amphibole) from a primitive asthenospheric melt. The restricted enrichment in HREE seen in melts may be attributed to involvement of garnet, which coexists with all the clinopyroxenes/melts represented in figure 4*a* except for the MARID-type ones. Even for the MARID series rocks there is a close association with garnetiferous peridotites (Dawson 1987; Erlank *et al.* 1987), so that there may well have been little differentiation of the MARID melt away from compositions in equilibrium with garnet. It is also significant to note that much of the overall spectrum of compositions shown in figure 4*a* is represented by two metasomatized rocks from Jagersfontein (J13, rich in clinopyroxene and mica; J134, with edenitic amphibole) where there is evidence of progressive evolution from amphibole-free to amphibole-bearing rocks (Winterburn *et al.* 1990).

Figure 4*b* shows a similar set of trace element abundance patterns for calculated melts in equilibrium with clinopyroxenes from alkali basalt sources. The data shown were chosen to encompass the broad spectrum of compositions found in Type II dykes, metasomatic rocks and megacrysts, but data from type Type IA spinel lherzolite xenoliths were excluded since these have been linked to residua depleted by partial melting (Frey & Prinz 1978; Menzies *et al.* 1985). Given the variety of sources, the data show a strikingly limited range of compositions. The most enriched liquid composition is that calculated from clinopyroxenes in a set of amphibole and apatite-rich xenoliths described by Wass *et al.* (1980) as fractionates. Slopes of the melt REE patterns vary widely, resulting in crossovers in the MREE range, but the steeper slopes (and Ce/Ho for example) are notably similar to those calculated for melt coexisting with the kimberlite-derived megacryst of figure 4*a*. Evidence for garnetiferous precursors to basalt-derived peridotites has been noted (Nicolas 1986), and the variations in LREE/HREE in figure 4*b* may be linked to the extent to which melt differentiation has occurred in the presence or absence of garnet (cf. Menzies *et al.* 1987, p. 347). Allowing for this factor, the spectrum of melt compositions in figure 4*b* shows many similarities to that in figure 4*a*. Thus derivation of all these melt

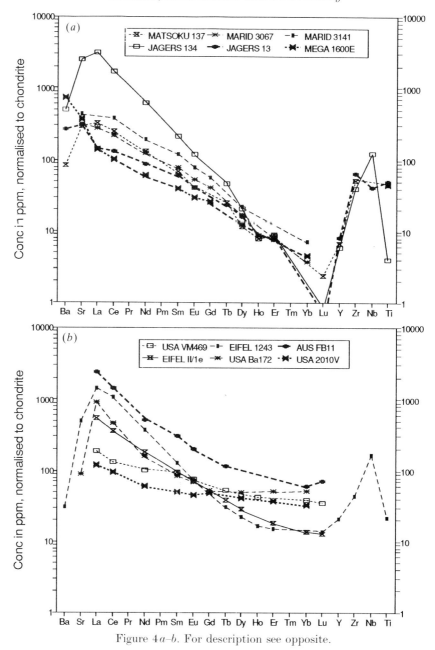

Figure 4 a–b. For description see opposite.

compositions from similar source liquids of primitive OIB-like trace element characteristics, would be possible by a differentiation process such as percolative fractional crystallization (as described above).

Exactly the same features are shown by the melt compositions calculated from amphiboles from various dykes, veins, metasomites and megacrysts (figure 4c). The melts calculated from the kimberlite-derived amphiboles have relatively high LREE/HREE, similar to some of the basalt-derived compositions, e.g. Eifel, while other basalt-derived amphiboles yielded melt compositions with lower LREE/HREE.

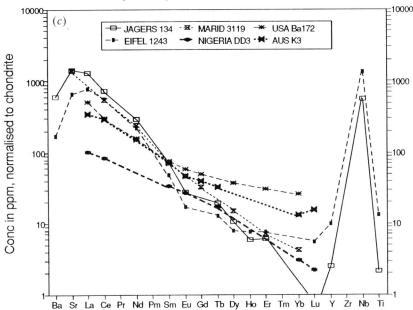

Figure 4. Melt compositions (chondrite normalized) coexisting with clinopyroxenes and amphiboles, calculated by using partition coefficients (see text). (*a*) Melts coexisting with clinopyroxenes from various kimberlite-derived metasomatic associations (Marid, Matsoku and Jagersfontein types; see table 2). In addition a Cr-poor megacryst composition has been used to calculate a melt composition (Mega 1600E) which is believed to be representative of primitive oib-like asthenospheric melts (Harte 1983; Jones 1987). (Matsoku and Jagersfontein data from this study, Marid and megacryst data from Kramers *et al.* (1981, 1983).) (*b*) Melts coexisting with clino-pyroxenes from basalt-derived Type II and IB xenoliths and megacrysts. (Eifel 1243 data this study, other clinopyroxene data from Wass *et al.* (1980), Irving & Frey (1984) and Menzies *et al.* (1985).) (*c*) Melts coexisting with amphibole. (The Jagersfontein, Marid and Eifel data are from this study, other source data from Irving & Frey (1984) and Menzies *et al.* (1985).)

However, the trace element abundances of melts calculated from amphibole compositions are generally lower than those calculated from clinopyroxenes, even for coexisting amphibole/clinopyroxene pairs in the same rock (e.g. Jagers 134, Eifel 1243, USA Ba172 in figure 4). The explanation is possibly that the amphibole partition coefficients used are too low by comparison with the clinopyroxene ones. Nevertheless, for our present purposes the important point is that the amphibole array of melt compositions again shows a spectrum of gradual changes such that all the melts may have arisen by differentiation processes (e.g. percolative fractional crystallization) under varying pressure–temperature conditions from a similar primitive basalt-picrite with oib-like trace element relative abundances.

5. Discussion

The trace element data presented above support the petrographic data (§3) in their evidence for close links between dyke, vein and metasomatic rocks, and in indicating that melt differentiation processes are an important aspect of the evolution of the various dyke, vein and modal metasomatic associations. But they go beyond this to further suggest that all such associations could result from differentiation of similar primitive melts with oib-like trace element compositions, occurring under a variety of pressure–temperature conditions.

Our major conclusion is therefore that the dykes/veins/metasomites in mantle xenoliths document processes of fractional crystallization and percolative fractional crystallization, by which primitive melts with OIB-like trace elements may differentiate to more evolved compositions, which in turn may lead to further metasomatism and enrichment in higher levels of incompatible elements. The crystallization of metasomatic and other minerals, whether it be in dykes, veins or metasomatised rocks, and their separation from the flowing/percolating melt provides the differentiation mechanism that leads to the formation of more extreme melts and their crystallization products. Thus the melts and their crystallization products evolve together, each begetting the other in the evolutionary process. This mutual relationship lies behind the conflicting arguments in the literature concerning whether metasomatism is precursory to or consequent upon magmatism (Bailey 1982; Menzies 1983; Wilshire 1987). Our thesis does not imply that the evolving melt composition eventually gives rise to the same melt (basaltic or kimberlitic) in which the mantle xenoliths are erupted; similar processes may have operated at widely separated intervals of time and thus the xenoliths may not be directly connected to the erupting melt.

Aspects of the similarities of melts/fluids involved in various types of modal metasomatism have been noted previously; e.g. Menzies *et al.* (1987) for a large set of trace element compositions, and Field *et al.* (1989) on the basis of amphibole compositions. However, unlike herein, these authors also placed emphasis on fluids distinct from melts. Menzies *et al.* (1987) considered that amphibole- and mica-rich metasomatism, particularly as represented by the richterite-bearing peridotites, was rather attributed to a hydrous fluid than a melt and emphasized low Ti characteristics as endorsing a hydrous origin. We suggest that such low-Ti characteristics may be a product of earlier and probably deeper stages of melt fractional crystallization in which abundant ilmenite is crystallized as in the Matsoku metasomites (Harte *et al.* 1987). Judging by the widespread occurrence of hydrous metasomatic phases, it is clear that the more evolved melts must be rich in H_2O, but the lack of any sharp distinction in the trace element abundance patterns of the hydrous phases contained in evolved metasomatic rocks in relation to the other assemblages (figure 4) provides no evidence for precipitation of the metasomatic phase from a separate hydrous fluid phase distinct from melt. A continuum of trace element compositions appears to exist, just as there is evidence of a continuum of petrographic characteristics (§3). This point has also been made by Kramers *et al.* (1983), and experimental data clearly support the possibility that appropriate hydrous assemblages can crystallize from a melt containing just 15% H_2O (Edgar *et al.* 1976; Waters 1987). This is not to say that a distinct hydrous phase cannot coexist with melt under appropriate circumstances, as shown by Odling & Randle (1992), but we believe that it is a differentiating melt phase which is the most likely source of the metasomatic phases.

Two other important points follow from the arguments that a spectrum of melt compositions, causing diverse aspects of modal metasomatism, can be derived by differentiation from asthenospheric melts with primitive OIB-like trace element characteristics. The first is that melts causing modal metasomatism may also be responsible for cryptic metasomatism (Dawson 1984), i.e. enrichment without modal or major-minor element change (Harte 1983, 1987). This style of metasomatism is recognized by trace element enriched (especially in LREE) compositions in clinopyroxenes in peridotites lacking petrographic evidence of metasomatism (e.g. the Type IB xenoliths from basalts, as classified by Menzies (1983)). Since the spectrum of enriched clinopyroxene compositions lies within the range of compositions seen in modal metasomatic peridotites (Neilsen & Noller 1987; Menzies *et al*. 1987) it should be possible to generate them by infiltration of the same set of melts that cause modal metasomatism. Thus they represent situations in which melt infiltrated the mantle peridotites and caused enriched trace element compositions in clinopyroxenes and other minerals, but did not cause the crystallization of any additional phases. This is fully compatible with the postulated addition of a second component to these rocks as suggested by Frey & Prinz (1978) and Frey (1980).

The second point concerns the origin of kimberlitic and lamproitic melts. Menzies *et al*. (1987), Erlank *et al*. (1987) and Waters (1987) show overlaps between the trace element compositions of kimberlites and lamproites, and various metasomatic and MARID rocks. The kimberlite/lamproite compositions fall within the compositional spectrum of melts calculated to be in equilibrium with dykes/veins/metasomites shown in figure 4. It follows that kimberlitic and lamproitic melts themselves might also be products of differentiation of primitive melts with OIB-like trace element composition, by processes of percolative fractional crystallization. In these cases, the percolation through and maintenance of local equilibrium with mantle peridotite, while undergoing crystal fractionation, is important in maintaining equilibrium with olivine and thereby ensuring the characteristic high levels of Ni in the melts. In effect the hypothesis put forward here concurs with the often-made suggestion that kimberlites are a differentiation product of the same magmas that crystallize typical megacrysts; but our proposal goes further in explaining the types of crystal fractionation and metasomatic interaction processes involved in this differentiation. However, we do not advocate that all kimberlites and lamproites arise directly by such differentiation. The differentiated liquids may well freeze in the mantle, giving rise to dykes/veins/metasomites, which upon subsequent melting, possibly much later, could form the kimberlitic and lamproitic melts (Waters 1987; McKenzie 1989). Such a two stage origin must be favoured for Group II kimberlites and those lamproites which show enriched Sr and Nd isotopic ratios. A one stage origin by differentiation from an OIB-like parent would only be appropriate for compositions like Group I kimberlites that have isotopic compositions falling within the normal mantle array indicated by basalts.

6. Conclusions

1. Surface energy considerations show that basic-ultrabasic melts should be dispersed along mineral grain edges to form a connected three-dimensional network even when present in very small volumes. This textural geometry coupled with viscosity data for the melts, indicates that very small volume melts can move through peridotitic matrices on relatively short geological timescales and enables melts to behave as metasomatic fluids.

2. Layered cumulates demonstrate the ability of basic-ultrabasic melt to flow through nearly wholly crystalline material, and precipitate mineral aggregates which do not locally represent the bulk composition of the melts.

3. The dykes, veins and modal metasomatic rocks recovered from the mantle show evidence of melt flow and accompanying crystallization processes, which leave mineral aggregates showing similarities to cumulates in textural and chemical characteristics, and which do not represent the bulk chemical composition of the

melt. The resultant rocks might be termed 'dyke cumulates' and 'metasomatic infill cumulates'.

4. The whole assemblage of minerals crystallized in any of the diverse and widely recognized associations of dykes, veins and modal metasomatic rocks, involves a wide array and continuum of silicate, oxide and sulphide minerals, and clearly argues for crystallization from silicate melts. No clear evidence for the operation of hydrous fluids distinct from melts has been found; with both modal mineralogical and trace element data indicating continuous ranges of compositions.

5. The formation of the various mantle dyke, veins, and metasomatic rocks demonstrates the occurrence of fractional crystallization of the melts forming them. Thus the melts undergo differentiation. These processes will often occur while the melt maintains equilibrium with the peridotitic matrix it is passing through; thus percolation (chromatographic column type) effects will also influence the evolution of melt and the term 'percolative fractional crystallization' is suggested for this combination of processes.

6. Calculated melt compositions in equilibrium with a wide array of dyke, vein and metasomatic clinopyroxene and amphibole compositions, indicate a related progressive series of compositions which might originate by fractionation or by percolative fractional crystallization from similar asthenospherically derived primitive melts with OIB-like trace element compositions. Such melt differentiation, to varying degrees under different pressure–temperature conditions and water contents, can account for the whole array of metasomatic fluid compositions encountered in xenoliths from both basalts and kimberlites. This applies not only to modal metasomatism but also to cryptic metasomatism (where there is no modal or other petrographic expression of the passage of melt).

7. The spectrum of melt compositions generated by percolative fractional crystallization, includes ones with similar trace element compositions to kimberlites and lamproites, and thereby provides a potential means of generating such melts.

We thank colleagues in Edinburgh, particularly John Craven, Richard Hinton and Nicholas Odling for help with ion microprobe analyses and discussion of the typescript. H. Seck and G. Witt-Eickschen provided specimens of Eifel xenoliths. Mike Cheadle's wisdom on textures has been much appreciated, and he, Martin Menzies and Brian Upton kindly reviewed the typescript. Angus Miller and Gavin Andrews helped greatly with data reduction and diagram production. NERC fund the Edinburgh Ion Microprobe Facility. Assistance from De Beers Consolidated Mines Ltd with finance and rock collection has also been important.

References

Bailey, D. K. 1982 Mantle metasomatism – continuing chemical change within the Earth. *Nature, Lond.* **296**, 525–530.

Beere, W. 1975 A unifying theory of the stability of penetrating liquid phases and sintering pores. *Acta metall.* **23**, 131–138.

Bulau, J. R., Waff, H. S. & Tyburczy, J. A. 1979 Mechanical and thermodynamic constraints on fluid distribution in partial melts. *J. geophys. Res.* **84**, 6201–6108.

Bulau, J. R. 1982 Intergranular fluid distribution in olivine-liquid basalt systems. Ph.D. dissertation, Yale University, U.S.A.

Cheadle, M. J. 1989 Properties of texturally equilibrated two phase aggregates. Ph.D. thesis, University of Cambridge.

Dawson, J. B. 1984 Contrasting types of upper mantle metasomatism. In *Kimberlites. II. The mantle and crust–mantle relationships* (ed. J. Kornprobst), pp. 289–294. Amsterdam: Elsevier.

Dawson, J. B. 1987 The MARID suite of xenoliths in kimberlite: relationship to veined and metasomatised peridotite xenoliths. In *Mantle xenoliths* (ed P. H. Nixon), pp. 465–473. Chichester: Wiley.

Dawson, J. B. & Smith, J. V. 1977 The MARID (mica-amphibole-rutile-ilmenite-diopside) suite of xenoliths in kimberlite. *Geochim. cosmochim. Acta* **41**, 309–323.

Edgar, A. D., Green, D. H. & Hibberson, W. O. 1976 Experimental petrology of a highly possanic magma. *J. Petrol.* **17**, 339–356.

Eggler, D. H. 1987 Solubility of major and trace elements in mantle metasomatic fluids: experimental constraints. In *Mantle metasomatism* (ed. M. A. Menzies & C. J. Hawkesworth), pp. 21–41. London: Academic Press.

Erlank, A. J., Waters, F. G., Hawkesworth, C. J., Haggerty, S. E., Allsopp, H. L., Rickard, R. S. & Menzies, M. 1987 Evidence for mantle metasomatism in peridotite nodules from the Kimberley pipes. South Africa. In *Mantle metasomatism* (ed. M. A. Menzies & C. J. Hawkesworth), pp. 221–309. London: Academic Press.

Field, S. W., Haggerty, S. E. & Erlank, A. J. 1989 Sub-continental metasomatism in the region of Jagersfontein, South Africa. In *Kimberlites and related rocks*, vol. 2, pp. 771–783. (Special Publication no. 14). Australia: Geological Society.

Frey, F. A. 1980 The origin of pyroxenites and garnet pyroxenites from Salt Lake Crater, Oahu, Hawaii, trace element evidence. *Am. J. Sci.* A **280**, 427–449.

Frey, F. A. & Prinz, M. 1978 Ultramafic inclusions for San Carlos Arizona: petrologic and geochemical data bearing on their petrogenesis. *Earth Planet. Sci. Lett.* **38**, 129–176.

Fujii, N., Osamura, K. & Takahashi, E. 1986 Effect of water saturation on the distribution of partial melt in the olivine-pyroxene-plagioclase system. *J. geophys. Res.* **91**, 9253–9259.

Griffin, W. L., Wass, S. Y. & Hollis, J. D. 1984 Ultramafic xenoliths from Bullermerri and Grotuk Maars, Victoria, Australia; petrology of a sub-continental crust–mantle transition. *J. Petrol.* **25**, 53–87.

Gurney, J. J. & Harte, B. 1980 Chemical variations in upper mantle nodules from southern African kimberlites. *Phil. Trans. R. Soc. Lond.* A **297**, 273–293.

Haggerty, S. E. 1983 The mineral chemistry of new titanates from the Jagersfontein kimberlite, South Africa: implications for metasomatism in the upper mantle. *Geochim. cosmochim. Acta* **47**, 1833–1854.

Harte, B. 1977 Rock nomenclature with particular relation to deformation and recrystallisation textures in olivine-bearing xenoliths. *J. Geology* **85**, 279–288.

Harte, B. 1983 Mantle peridotites and processes – the kimberlite sample. In *Continental basalts and mantle xenoliths* (ed. C. J. Hawkesworth & M. J. Norry), pp. 46–91. Shiva.

Harte, B. 1987 Metasomatic events recorded in mantle xenoliths: an overview. In *Mantle xenoliths* (ed. P. H. Nixon), pp. 625–640. Chichester: Wiley.

Harte, B. & Hawkesworth, C. J. 1989 Mantle domains and mantle xenoliths. In *Kimberlites and related rocks*, vol. 2, 649–686 (Special Publication no. 14). Geological Society of Australia.

Harte, B. & Hunter, R. H. 1986 Speculations concerning the importance of metasomatic melt migration in the formation of pyroxenite sheets in garnet peridotite xenoliths from Matsoku, Lesotho. In *International Kimberlite Conference, Extended Abstracts*, pp. 184–186. Australia: Geological Society (Abstract Series no. 16).

Harte, B. & Matthews, M. B. 1989 Melt textures and ilmenite concentrations in mantle 'dykes' and metasomatic rocks. *Extended Abstracts, 28th, Int. Geol. Congress* **2**, 32–33.

Harte, B., Winterburn, P. A. & Gurney, J. J. 1987 Metasomatic phenomena in garnet peridotite facies mantle xenoliths from the Matsoku kimberlite pipe, Lesotho. In *Mantle metasomatism* (ed. M. A. Menzies & C. J. Hawkesworth), pp. 145–220. London: Academic Press.

Harte, B., Pattison, D. R. M. & Linklater, C. M. 1991a Field relations and petrography of partially melted pelitic and semi-pelitic rocks. In *Equilibrium and kinetics in contact metamorphism* (ed. G. Voll, J. Topel, D. R. M. Pattison & F. Seifert), pp. 181–209. Berlin: Springer-Verlag.

Harte, B., Matthews, M. B., Winterburn, P. A. & Gurney, J. J. 1991b Aspects of melt composition, crystallization, metasomatism and distribution, shown by mantle xenoliths from the Matsoku Kimberlite pipe. In *Fifth International Kimberlite Conference Extended Abstracts: CPRM (Special Publication 2/19*, pp. 167–169. Brasilia.

Hofmann, W. W. 1972 Chromatographic theory of infiltration metasomatism and its application to feldspars. *Am. J. Sci.* **272**, 69–90.

Hunter, R. H. 1987 Textural equilibrium in layered igneous rocks. In *Origins of igneous layering* (ed. I. Parsons), pp. 473–503. D. Reidel.

Hunter, R. H. & McKenzie, D. 1989 The equilibrium geometry of carbonate melts in rocks of mantle composition. *Earth planet. Sci. Lett.* **92**, 347–356.

Irving, A. J. 1980 Petrology and geochemistry of composite ultramafic xenoliths in alkalic basalts and implications for magmatic processes within the mantle. *Am. J. Sci.* A **280**, 389–426.

Irving, A. J. & Frey, F. A. 1984 Trace element abundances in megacrysts and their host basalts: constraints on partition coefficient and megacryst genesis. *Geochim. cosmochim. Acta* **48**, 1201–1221.

Jones, R. A. 1987 Strontium and neodymium isotopic and rare earth element evidence for the genesis of megacrysts in kimberlites of southern Africa. In *Mantle xenoliths* (ed. P. H. Nixon), pp. 711–724. Chichester: Wiley.

Jones, A. P., Smith, J. V. & Dawson, J. B. 1982 Mantle metasomatism in 14 veined peridotites from Bultfontein mine, South Africa. *J. Geology* **90**, 135–153.

Jurewicz, S. R. & Jurewicz, A. J. G. 1986 Distribution of apparent angles on random sections with emphasis on dihedral angle measurements. *J. geophys. Res.* **91**, 9277–9282.

Jurewicz, S. R. & Watson, E. B. 1985 Distribution of partial melts in granitic system: the application of liquid phase sintering. *Geochim. cosmochim. Acta* **49**, 1109–1121.

Kingery, W. D., Bowen, H. K. & Uhlmann, D. R. 1976 *Introduction to ceramics.* New York: Wiley Interscience.

Kramers, J. D., Smith, C. B., Lock, N. P., Harmon, R. S. & Boyd, F. R. 1981 Can kimberlites be generated from an ordinary mantle. *Nature, Lond.* **291**, 53–56.

Kramers, J. D., Roddick, J. C. M. & Dawson, J. B. 1983 Trace element and isotopic studies on veined, metasomatic and 'MARID' xenoliths from Bultfontein, South Africa. *Earth planet. Sci. Lett.* **65**, 90–106.

Laporte, D. 1988 Wetting angle between silicic melts and biotite (abstract). *Eos, Wash.* **69**, 1411.

Lloyd, F. E. 1987 Characterization of mantle metasomatic fluids in spinel cherzolites and alkali clinopyroxenites from the west Eifel and south west Uganda. In *Mantle metasomatism* (ed. M. A. Menzies & C. J. Hawkesworth), pp. 91–123. London: Academic Press.

Matthews, M. B. & Harte, B. 1989 Preservation of melt textures in mantle xenoliths. *Terra Abstr.* **1**, 274.

Matthews, M., Harte, B. & Prior, D. 1992 Mantle garnets: a cracking yarn. *Geochim. cosmochim. Acta* **56**, 2633–2642.

McKenzie, D. P. 1984 The generation and compaction of partially molten rock. *J. Petrol.* **25**, 713–765.

McKenzie, D. 1989 Some remarks on the movement of small melt fractions in the mantle. *Earth planet. Sci. Lett.* **95**, 53–72.

McKenzie, D. & O'Nions, R. K. 1991 Partial melt distributions from inversion of rare earth concentrations. *J. Petrol.* **32**, 1021–1091.

Menzies, M. A. 1983 Mantle ultramafic xenoliths in alkaline magmas: evidence for mantle heterogeneity modified by magmatic activity In *Continental basalts and mantle xenoliths* (ed. C. J. Hawkesworth & M. J. Norry), pp. 92–110, Shiva: Nantwich.

Menzies, M. A., Kempton, P. & Dungon, M. 1985 Interaction of continental lithosphere and asthenosphere melts below the Geronimo volcanic field. Arizona, U.S.A. *J. Petrol.* **26**, 663–693.

Menzies, M. A., Rogers, N., Tindle, A. & Hawkesworth, C. J. 1987 Metasomatic and enrichment processes in lithospheric peridolites, an affect of asthenosphere–lithosphere interaction. In *Mantle metasomatism* (ed. M. Menzies & C. J. Hawkesworth), pp. 313–361. London: Academic Press.

Navon, O. & Stolper, E. 1987 Geochemical consequences of melt percolation: the upper mantle as a chromatographic column. *J. Geology* **95**, 285–307.

Nicholas, A. 1986 A melt extraction model based on structural studies in mantle peridotites. *J. Petrol.* **27**, 999–1022.

Nicholas, A. & Jackson, M. 1982 High temperature dikes in peridolites: origin by hydraulic fracturing. *J. Petrol.* **23**, 568–582.

Nielson, J. E. & Noller, J. S. 1987 Processes of mantle metasomatism; constraints from observations of composite peridotite xenoliths. *Geol. Soc. Am. Spec. Pap.* **215**, 61–76.

Odling, N. W. A. & Randle, H. A. 1992 The partitioning of trace elements between fluid and melt. *Terra Nova (abstracts)* **4**, 32–33.

O'Nions, R. K. & McKenzie, D. P. 1988 Melting and continent generation. *Earth planet. Sci. Lett.* **90**, 449–456.

Riley, G. N. & Kohlstedt, D. L. 1991 Kinetics of melt migration in upper mantle rocks. *Earth planet. Sci. Lett.* **105**, 500–521.

Smith, C. S. 1948 Grains phases and interfaces: an interpretation of microstructure. *Trans AIME* **197**, 15–51.

Sun, S.-S. & MacDonough, W. F. 1989 Chemical and isotopic systematics of ocean basalts; implications for mantle composition and processes. In *Magmatism in the ocean basins* (ed. A. D. Saunders & M. J. Norry), pp. 313–345. *Geol. Soc. Lond. Spec. Publ.* **42**.

Toramaru, A. & Fujii, N. 1986 Connectivity of a melt phase in a partially molten peridotite. *J. geophys. Res.* **91**, 9239–9252.

Vincenzi, E. P., Rapp, R. & Watson, E. B. 1988 Crystal/melt wetting characteristics in partially molten amphibolite (abstract). *Eos, Wash.* **69**, 482.

Voll, G. 1960 New work on petrofabrics. *Liverp. Manch. Geol. J.* **1**, 73–85.

Von Bargen, N. & Waff, H. S. 1988 Wetting of enstatite by basaltic melt at 1350 °C and 1.0–2.5 GPa pressure. *J. geophys. Res.* **93**, 1153–1158.

Waff, H. S. & Bulau, J. R. 1979 Equilibrium fluid distribution in an ultramafic partial melt under hydrostatic stress conditions. *J. geophys. Res.* **84**, 6109–6114.

Wager, L. R., Brown, G. M. & Wadsworth, W. J. 1960 Types of igneous cumulates. *J. Petrol.* **1**, 73–85.

Walker, D., Jurewicz, S. & Watson, E. B. 1988 Adcumulus dunite growth in a laboratory thermal gradient. *Contrib. Mineral. Petrol.* **99**, 306–319.

Wass, S. Y., Henderson, P. & Elliot, C. J. 1980 Chemical heterogeneity and metasomatism in the upper mantle: evidence from rare earth and other elements in apatite rich xenoliths in basaltic rocks from eastern Australia. *Phil. Trans. R. Soc. Lond.* A **297**, 333–346.

Waters, F. G. 1987 A suggested origin of MARID xenoliths in kimberlites by high pressure crystallisation of an ultrapotassic rock such as lamproite. *Contrib. Miner. Petrol.* **95**, 523–533.

Watson, E. B. & Brenan, J. M. 1987 Fluids in the lithosphere. I. Experimentally determined wetting characteristics of CO_2–H_2O fluids and their implications for fluid transport, host-rock physical properties and fluid inclusion formation. *Earth planet. Sci. Lett.* **85**, 497–515.

Wilshire, H. G. 1987 A model of mantle metasomatism. *Geol. Soc. Am. Spec. Vol.* **215**, 47–60.

Wilshire, H. G. & Pike, J. E. N. 1975 Upper mantle diapirism: evidence from analogous features in alpine peridotite and ultramafic inclusions in basalt. *Geology* **3**, 467–470.

Wilshire, H. G. & Shervais, J. W. 1975 Al augite and Cr diopside ultramafic xenoliths in basaltic rocks from the western United States. *Phys. Chem. Earth* **9**, 257–272.

Wilshire, H. G. & Trask, N. J. 1971 Structural and textural relationships of amphibole and phlogopite in peridotite inclusions, Dish Hill, California. *Am. Mineralogist* **56**, 240–251.

Wilshire, H. G., Pike, J. E. N., Meyer, C. E. & Schwarzmann, E. C. 1980 Amphibole-rich veins in lherzolite xenoliths, Dish Hill and Deadman Lake, California. *Am. J. Sci.* A **280**, 576–593.

Witt, G. & Seck, H. A. 1987 Temperature history of sheared mantle xenoliths from west Eifel, west Germany: evidence for mantle diapirism beneath the Rhenish Massif. *J. Petrol.* **28**, 475–494.

Winterburn, P. A., Harte, B. & Gurney, J. J. 1990 Peridotite xenoliths from the Jagersfontein kimberlite pipe. I. Primary and primary-metasomatic mineralogy. *Geochim. cosmochim. Acta* **54**, 329–341.

Wyatt, B. A. & Lawless, P. J. 1984 Ilmenite in polymict xenoliths from the Bultfontein and De Beers mines, South Africa. In *Kimberlites II: the mantle and crust–mantle relationships* (ed. J. Kornprobst), pp. 43–56. Amsterdam: Elsevier.

Physics of melt extraction: theory, implications and applications

By Marc Spiegelman

Lamont-Doherty Geological Observatory, of Columbia University, Palisades, New York 10964, U.S.A.

This paper presents a general overview of flow in deformable porous media with emphasis on melt extraction processes beneath mid-ocean ridges. Using a series of simple model problems, we show that the equations governing magma migration have two fundamentally different modes of behaviour. Compressible two-phase flow governs the separation of melt from the solid and forms a nonlinear wave equation that allows melt to propagate in solitary waves. Incompressible two-phase flow governs small-scale mantle convection driven by lateral variations in melt content. The behaviour of both compressible and incompressible matrix deformation is demonstrated in the context of mid-ocean ridges to show that both mechanisms may explain the observation of the narrowness of ridge volcanism. These results also suggest that melt extraction is an inherently time dependent process that may account for the timing, volume and chemistry of volcanism.

1. Introduction

Magma migration and other important geophysical two-phase flows have long been considered to be efficient processes for transporting heat and mass in the mantle and the crust. However, it is only with the derivation of a comprehensive system of conservation equations (McKenzie 1984; Scott & Stevenson 1984, 1986; Fowler 1985) that we have begun to understand the fluid mechanics of these processes. The purpose of this paper is to develop a better physical understanding of the behaviour of two-phase flows through a series of simple model problems. Section 2a reviews the general equations governing viscous two-phase flows and presents some model problems to demonstrate that magma migration is an inherently time-dependent process. Section 3 then illustrates the behaviour of the governing equations in a more geological context to understand some of the possible processes involved in melt extraction at mid-ocean ridges. Given a better understanding of the general physics, the geological implications of these problems becomes clear and we can begin to use this insight to design experiments to test the theory and use observations to reveal the properties of the partly molten mantle.

2. General physics

(a) Governing equations

The equations governing the percolative flow of a low viscosity fluid or 'melt' through a viscously deformable permeable matrix were derived independently by several workers (McKenzie 1984; Scott & Stevenson 1984, 1986; Fowler 1985) based

Phil. Trans. R. Soc. Lond. A (1993) **342**, 23–41

Printed in Great Britain

on the more general work of Drew (1971, 1983) for interpenetrating two-phase flows. Of these formulations, McKenzie (1984) provides a particularly detailed derivation and is perhaps the more general. These equations are a macroscopic description of two interpenetrating viscous fluids with vastly different viscosities. The melt is assumed to form an interconnected porous network distributed over some characteristic pore (or vein spacing) a. As a continuum approximation, these equations are valid for length scales much larger than a and smaller than any characteristic variation in porosity. These equations also assume that inertial effects are negligible for both the percolating melt phase and for creeping matrix deformation. This assumption should remain valid as long as the porosity is much smaller than the critical value at which the matrix disaggregates.

With these considerations, the equations governing conservation of mass and momentum can be written

$$\partial(\rho_{\mathrm{f}}\phi)/\partial t + \nabla\cdot(\rho_{\mathrm{f}}\phi\boldsymbol{v}) = \Gamma, \tag{1}$$

$$\partial[\rho_{\mathrm{s}}(1-\phi)]/\partial t + \nabla\cdot[\rho_{\mathrm{s}}(1-\phi)\,\boldsymbol{V}] = -\Gamma, \tag{2}$$

$$\phi(\boldsymbol{v}-\boldsymbol{V}) = (-k_{\phi}/\mu)\,\nabla\mathscr{P}, \tag{3}$$

$$\frac{\partial\mathscr{P}}{\partial x_i} = \frac{\partial}{\partial x_j}\eta\left(\frac{\partial V_i}{\partial x_j}+\frac{\partial V_j}{\partial x_i}\right)+\frac{\partial}{\partial x_i}(\zeta-\tfrac{2}{3}\eta)\,\nabla\cdot\boldsymbol{V}-(1-\phi)\,\Delta\rho g\delta_{i3}. \tag{4}$$

$$k_{\phi} \sim a^2\phi^n/b. \tag{5}$$

In (1)–(5), ρ_{f} is the density of the melt, ϕ is the volume fraction occupied by the melt or porosity, $\boldsymbol{v}$ is the melt velocity, and Γ is the rate of mass transfer from matrix to melt (melting rate). ρ_{s} is the density of the solid matrix, $\boldsymbol{V}$ is the matrix velocity, k_{ϕ} is the permeability, μ is the melt viscosity and $\mathscr{P} = P-\rho_{\mathrm{f}}gz$ is the pressure in excess of hydrostatic pressure. $\mathscr{P}$ is often referred to as the 'piezometric pressure', or 'hydraulic head'. η and ζ are respectively, the matrix shear and bulk viscosities and $\Delta\rho = \rho_{\mathrm{s}}-\rho_{\mathrm{f}}$. Estimates and explanations of additional parameters are given in table 1.

Equations (1) and (2) conserve mass for the melt and matrix individually but allow mass transfer between solid and liquid via the melting rate Γ. If Γ is positive the solid melts. $\Gamma < 0$ implies freezing. Equation (3) is a modified form of Darcy's Law which governs the separation of melt from the matrix. The separation flux of liquid from solid $\phi(\boldsymbol{v}-\boldsymbol{V})$ is proportional to the permeability and flows down pressure gradients. Equation (4) is Stokes equation for creeping matrix flow and shows that the pressure gradients that make melt move depend both on the buoyancy difference between melt and solid and on the viscous deformation of the matrix. Equation (5) gives the permeability as a nonlinear scalar function of the pore spacing a, porosity and a dimensionless coefficient, b. Equation (5) is a convenient parametrization for a range of porosity/permeability relationships valid for small porosities ($\leqslant 10$–20%). More specific relations can be found in standard texts on porous flow (Dullien 1979; Scheidegger 1974; Bear 1988). The actual functional form of the permeability is not particularly crucial to the following discussion except for the important requirements that both the permeability, k_{ϕ}, and $\partial k_{\phi}/\partial\phi$ are increasing functions of porosity. Simple capillaric models for permeability suggest that a power law with $n \approx 2$–3 is a good approximation for natural systems. More detailed analysis of texturally equilibrated melt/solid networks gives similar results (Cheadle 1989; Von Bargen & Waff 1986).

Table 1. *Notation*

variable	meaning	value used	dimension
a	pore spacing (grain size)	$10^{-3}-10^{-1}$	m
b	constant in permeability	100–3000	none
$\mathscr{C}$	compaction rate (isotropic strain rate)	—	s^{-1}
g	acceleration due to gravity	9.81	$m\ s^{-2}$
k_ϕ	permeability	—	m^2
k_0	$= a^2\phi_0^n/b$ permeability at porosity ϕ_0	—	m^2
n	exponent in permeability	2–3	none
$\mathscr{P}$	$= P - \rho_{\rm f} gz$ piezometric pressure	—	Pa
q	$= \phi v$ dimensionless melt flux	—	none
t	time	—	s
$\mathscr{U}^{\rm s}$	matrix scalar potential	—	$m^2\ s^{-1}$
V	$= \nabla \times \boldsymbol{\varPsi}^{\rm s} + \nabla \mathscr{U}^{\rm s}$ matrix velocity	—	$m\ s^{-1}$
v	melt velocity	—	$m\ s^{-1}$
w_0	$= k_0 \Delta\rho g/\phi_0 \mu$ percolation velocity	—	$m\ s^{-1}$
x	horizontal cartesian coordinate	—	m
z	vertical cartesian coordinate	—	m
δ	$= \sqrt{(k_0(\zeta + \frac{4}{3}\eta)/\mu)}$ compaction length	100–10000	m
$\varGamma$	melting rate	—	$kg\ m^{-3}\ s^{-1}$
ζ	matrix bulk viscosity	$10^{18}-10^{21}$	Pa s
η	matrix shear viscosity	$10^{18}-10^{21}$	Pa s
ξ	$\eta/(\zeta + \frac{4}{3}\eta)$	$\frac{3}{7}$	none
ξ'	ξ/ϕ_0	—	none
μ	melt shear viscosity	1–10	Pa s
$\rho_{\rm f}$	density of melt	2800	$kg\ m^{-3}$
$\rho_{\rm s}$	density of matrix	3300	$kg\ m^{-3}$
$\Delta\rho$	$= \rho_{\rm s} - \rho_{\rm f}$	500	$kg\ m^{-3}$
ϕ	porosity	—	none
ϕ_0	reference porosity (constant)	0.005–0.04	none
$\psi^{\rm s}$	matrix stream function (2D)	—	$m^2\ s^{-1}$
$\boldsymbol{\varPsi}^{\rm s}$	matrix 3D vector potential	—	$m^2\ s^{-1}$

(b) *Governing equations in potential form: constant viscosity, constant densities*

The governing equations (1)–(5) are similar to those for standard porous media flow with the important distinction that the matrix is able to deform viscously. For the problem of magma migration from the mantle, this behaviour must be allowed if these equations are to be consistent with mantle convection. It is the pressure gradients due to this additional viscous deformation that provide for most of the interesting new behaviour. For the case of constant matrix viscosities, (4) can be written

$$\nabla\mathscr{P} = -\eta\nabla \times \nabla \times V + (\zeta + \tfrac{4}{3}\eta)\nabla(\nabla \cdot V) - (1-\phi)\Delta\rho g\boldsymbol{k} \qquad (6)$$

to show that the pressure gradients due to viscous flow arise from two fundamentally different types of matrix deformation. The first term of (6) is the pressure gradient due to incompressible shear (rotational flow) of the viscous matrix. The second term is the gradient due to volume changes of the compressible matrix. The distinction between compressible and incompressible matrix deformation provides a convenient way to categorize the general behaviour of the governing equations. It should be stressed that 'compressible' matrix deformation refers to the ability of the matrix framework to expand and compact to produce changes in porosity in response to variations in the melt flux. The individual crystals that form this framework, however, are incompressible. Incompressible flow refers to matrix shear without volume changes. By assuming constant viscosities, the pressure gradients due to the two basic modes of matrix deformation separate. Using Helmholtz's theorem, the matrix velocity field V can also be decomposed into incompressible and compressible flow fields:

$$V = \nabla \times \boldsymbol{\Psi}^s + \nabla \mathcal{U}^s, \tag{7}$$

$\boldsymbol{\Psi}^s$ is the vector potential governing rotational flows (in 2D $\boldsymbol{\Psi}^s$ is the stream function). $\mathcal{U}^s$ is a scalar potential governing irrotational flow. It is also useful to define the isotropic strain rate or 'compaction rate' as

$$\mathcal{C} = \nabla \cdot V. \tag{8}$$

The compaction rate is simply the rate of volume change of the matrix. If $\mathcal{C} > 0$ the matrix is expanding; $\mathcal{C} < 0$ implies compaction. Given these definitions and letting ρ_f, and ρ_s be constant, (1)–(5) can be rewritten solely in terms of porosity and matrix deformation. Expanding (2) gives

$$\partial \psi / \partial t + (\nabla \times \boldsymbol{\Psi}^s + \nabla \mathcal{U}^s) \cdot \nabla \psi - (1 - \psi) \mathcal{C} + \Gamma / \rho_s. \tag{9}$$

Adding (1) and (2) and substituting (3) and (6) yields

$$-\nabla \cdot (k_\phi / \mu) (\zeta + \tfrac{4}{3}\eta) \nabla \mathcal{C} + \mathcal{C} = \nabla \cdot (k_\phi / \mu) [\eta \nabla \times \nabla^2 \boldsymbol{\Psi}^s - (1 - \phi) \Delta \rho g \boldsymbol{k}] + \Gamma \Delta \rho / \rho_s \rho_f. \tag{10}$$

Equations (7) and (8) imply

$$\nabla^2 \mathcal{U}^s = \mathcal{C} \tag{11}$$

and taking the curl of (6) gives

$$\nabla^4 \boldsymbol{\Psi}^s = -(\Delta \rho g / \eta) \nabla \times \phi \boldsymbol{k}. \tag{12}$$

Further details of the derivation, non-dimensionalization and a discussion of boundary conditions can be found in Spiegelman (1993*a*).

Equations (9)–(12) form a coupled system of hyperbolic, elliptic and bi-harmonic equations that can be solved by using standard techniques. While writing the equations in potential form may appear to obscure the physics, this decomposition actually makes the behaviour of viscous two-phase flow clearer by making the equations for incompressible and compressible matrix deformation explicit. This decomposition also allows these two modes to be solved for sequentially.

Equation (12) governs incompressible matrix deformation and is the Stokes equation for the creeping rotational flow of the matrix driven by horizontal porosity variations and boundary conditions. This equation is identical to that found in thermal convection. Here, however, the matrix convection is driven by variations in the amount of melt present rather than by variations in temperature. Considerable work has been carried out on incompressible two-phase shear flow, particularly at mid-ocean ridges (Rabinowicz *et al.* 1984; Spiegelman & McKenzie 1987; Phipps Morgan 1987; Ribe & Smooke 1987; Ribe 1988; Scott 1988; Scott & Stevenson 1989; Buck & Su 1989; Daly & Richter 1989; Cordery & Phipps Morgan 1992). Section 3*a* summarizes this work, and discusses its role in focusing melt movement beneath mid-ocean ridges.

Equations (9)–(11) govern compressible matrix deformation and have been less extensively studied. These equations can be combined to form a nonlinear dispersive wave equation for the evolution of porosity in space and time. The existence of nonlinear solitary wave solutions for porosity were demonstrated almost as soon as the equations were derived (Scott & Stevenson 1984; Scott *et al.* 1986; Richter & McKenzie 1984; Barcilon & Richter 1986). These waves have been shown to exist in one, two and three dimensions and their stability has been discussed by Scott & Stevenson (1986) and by Barcilon & Lovera (1989). While this previous work shows that solitary waves are easily produced, their overall significance has not been well understood. The more general physics of compressible matrix deformation is presented in Spiegelman (1993 a, b) which shows that these solitary waves are an essential feature of the governing equations and will form spontaneously from any perturbation in the melt flux. This work also investigates the effects of melting and freezing on the solitary waves and shows that many geologically reasonable initial conditions should develop solitary waves. The following section briefly reviews the behaviour of compressible flow that is important for understanding the behaviour of high permeability melt channels at ridges (§3 b).

(c) General behaviour of compressible flow

The important behaviour of compressible matrix deformation is contained in (9) and (10). Equation (9) governs conservation of porosity and states that changes in ϕ are caused by matrix advection and the balance between volume changes of the matrix and melting. Equation (10) governs volume changes of the matrix and states that the compaction rate $\mathscr{C}$ depends on the divergence of the melt separation flux $\phi(v-V)$ and the volume change on melting, i.e.

$$\mathscr{C} = -\nabla \cdot \phi(v-V) + \Gamma \Delta \rho / \rho_s \rho_f. \tag{13}$$

Equation (10), however, has been rearranged to stress that it is an elliptic equation for the compaction rate. The right side of (10) contains volume changes due to spatial variations in the 'forced flux' and to volume changes on mass transfer. The forced flux contains the components of the melt separation flux driven by incompressible shear and by buoyancy. The left side of (10) shows that these forcing terms change the volume of the matrix (second term) and develop an opposing 'compaction flux'. The compaction flux arises from pressure gradients induced by the viscous resistance of the matrix to changing volume.

The first term (10) is perhaps the most important term in the governing equations. Dimensional analysis shows that pressure gradients due to volume changes of the viscous matrix only become significant when the melt flux varies over the compaction length

$$\delta = [k_\phi(\zeta + \tfrac{4}{3}\eta)/\mu]^{\frac{1}{2}}. \tag{14}$$

In many geological problems, δ is small (order 100–1000 m) and several authors (Ribe 1985; Ribe & Smooke 1987; Scott & Stevenson 1989) have proposed that the first term in (10) can be neglected for most geological problems. Other workers (Buck & Su 1989; Sotin & Parmentier 1989) have simply neglected this term altogether. The zero compaction length approximation reduces a potentially singular second order equation to a zero-order equation. In particular, Spiegelman (1993 a) shows that the zero compaction length approximation reduces the equations for compressible flow to a single nonlinear wave equation that predicts porosity 'shock waves', travelling discontinuities in the porosity. Such shocks will result from any initial condition where the melt flux locally exceeds what can be readily extracted, for example, the injection of magma into a region of low permeability or a local increase in the melting rate. The shaded curves in figure 1 show the evolution of a simple initial condition

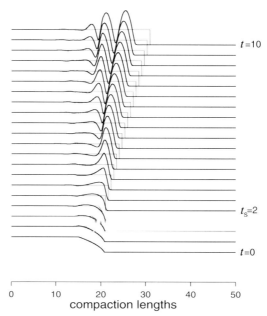

Figure 1. A comparison of approximate and full solutions for the evolution of porosity from an initial condition that develops into a perfect step function shock at time $t_s = 2$ in the zero compaction length approximation. The shaded lines show the analytic shock solution to the approximate equations. After $t = t_s$, this shock travels as a perfect step with $\phi_{max} = 1$, $\phi_1 = 0.2$ at constant velocity $c_s = 1.24$. The solid lines show the numerical solution to the full equations for the same initial condition. Until $t = t_s$ the full and approximate solutions are comparable. For $t \geqslant t_s$, viscous resistance of the matrix to volume changes causes the shock to disperse into a series of porosity maxima and minima. In this example, the leading porosity wave travels slower than the shock ($c \approx 0.8$) and each new wave forms further back relative to the matrix.

that forms a travelling step-function shock when the viscous resistance of the matrix is neglected. In the vicinity of the shock, however, viscous resistance to volume changes cannot be negligible and the approximation is not uniformally valid. The solid curves in figure 1 show the evolution of the identical initial condition using the full equations and shows that viscous resistance to volume changes causes the shock to disperse into a growing train of solitary waves. Figure 2 shows the long term evolution of porosity for a set of initial conditions that would produce simple shocks in the zero compaction length approximation. This figure shows that the amplitude of the solitary waves correlates with the size of the obstruction in flux (large steps produce large solitary waves) and that the dispersive wave trains can grow to span many compaction lengths and even allow for information to propagate backwards relative to the matrix (see figure 2 for $\phi_1 = 0.2$). Spiegelman (1993b) discusses and quantifies this dispersion and shows it to be consistent with the form of the governing equations. This work also shows that solitary waves are waves of volume fraction (porosity) and move faster than does the melt itself.

The solutions shown in figures 1 and 2 demonstrate the important processes that govern the separation of melt from solid. In general, porosity waves are a fundamental feature of these equations. These waves propagate because variations in the melt flux force the matrix to change volume. As long as the flux is an increasing function of porosity and the matrix is deformable, then porosity waves will exist. The actual speed and behaviour of the porosity waves, however, depends on the

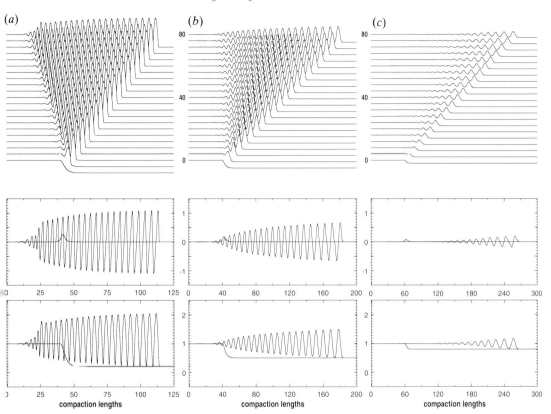

Figure 2. The long-term evolution of an initial condition similar to that in figure 1 showing the relationship between the amplitude of solitary waves and the size of the initial obstruction. In the absence of viscous dispersion, this initial condition would evolve to a travelling step function shock with $\phi_{max} = 1$, $\phi_{min} = \phi_1$. Viscous resistance to volume changes causes all shocks to disperse into rank-ordered dispersive wave trains of solitary waves. For each value of ϕ_1, the top figure shows porosity profiles against time ($t = 0$–80). Bottom two figures superpose initial ($t = 0$) and final ($t = 80$) profiles for compaction rate and porosity. All grids have $4\phi_1^{-\frac{3}{2}}$ grid points/compaction length for accurate resolution (Spiegelman 1993b). (a) A large-amplitude change in porosity ($\phi_1 = 0.2$) develops into a large-amplitude, slow moving, back-propagating wave train. (b) A smaller obstruction ($\phi_1 = 0.5$) develops a smaller-amplitude, faster wave train (note the difference in scales) with all new waves initiating at $ca.\ z = 40$. (c) The smallest obstruction ($\phi_1 = 0.8$) produces a small-amplitude, fast moving, forward propagating, wave train.

relationship between permeability and porosity. For most commonly used permeability–porosity relationships, $\partial k_\phi / \partial \phi$ is an increasing function of porosity; therefore, regions of high porosity (high melt flux) will propagate faster than regions of lower porosity and can cause the porosity gradients to steepen. If there was no viscous resistance of the matrix to changing volume, this steepening would continue until shocks form. However, in the vicinity of a rapid change in flux, viscous resistance cannot be negligible and in general it will cause the porosity to disperse into nonlinear solitary waves.

It should be stressed that because porosity gradients can steepen, viscous effects can become important even for smooth initial conditions where compaction effects are initially negligible. Spiegelman (1993a) presents the general criteria for shock formation using the zero compaction length approximation. Figures 1 and 2 also show that, once viscous effects become important, the obstruction in flux does not remain localized, but rather, excites the growth of additional solitary waves. Because of this dispersion, even initially localized disturbances in the melt flux can eventually affect the entire partly molten region. Thus a small compaction length does not imply a negligible compaction term. If these equations are an accurate description of magma migration in the mantle, the clear implication is that the episodicity of magmatism may reflect the basic processes of migration from the source region.

3. Using the theory: applications to mid-ocean ridges

The simple model problems of the previous section demonstrate the important behaviour of compressible matrix deformation, but do so in a somewhat abstract manner. The purpose of the following section is to develop a better understanding of the behaviour of the governing equations in a more geological context. Here we will use the theory of viscous two-phase flows to consider some of the processes that may be occurring at mid-ocean ridge spreading centres. In particular, this section is motivated by the important observation of the extreme narrowness of the neo-volcanic zone. As many authors have noted (MacDonald 1982; Detrick *et al.* 1987; Burnett *et al.* 1989; Toomey *et al.* 1990; Caress *et al.* 1992) a variety of geophysical data strongly suggest that the entire 6 km thickness of oceanic crust is emplaced within approximately 2 km of the ridge axis. The implications of this observation are clear. Either the melting region at depth is of a comparable width, or there must be some mechanism for lateral migration of melt to the ridge axis. The following sections will demonstrate three different mechanisms that can produce focused volcanism. The first two mechanisms depend on incompressible matrix shear. The third mechanism relies upon the behaviour of compressible matrix deformation. The three mechanisms illustrate the broad range of behaviour inherent in the equations for flow in deformable porous media and suggest that both incompressible and compressible matrix deformation may be significant in magma migration.

(a) *Incompressible flow: focusing by mantle shear*

The first two mechanisms for focused ridge volcanism rely on the ability of the matrix to shear. At mid-ocean ridges there are always two primary sources of mantle shear. The first is shear driven by boundary conditions, in this instance the spreading plates. The second is mantle shear driven by internal buoyancy variations. By rescaling (12) using the scaling relations in table 2, the dimensionless 2D equations for incompressible porosity driven convection can be written

$$\nabla^4 \psi^s = R \, \partial\phi/\partial x, \tag{15}$$

where $R = \phi_0 \Delta\rho g d^2/\eta U_0$ is the single parameter which measures the relative contributions of buoyancy driven shear to plate driven shear. Here d is the depth of the partly molten layer (*ca.* 50–60 km), U_0 is the half-spreading rate, and ψ^s is the 2D matrix stream function. Comparison with the dimensionless equations for thermal convection

$$\partial T/\partial t + (\nabla \times \psi^s \boldsymbol{j}) \cdot \nabla T = \nabla^2 T, \tag{16}$$

$$\nabla^4 \psi^s = Ra \, \partial T/\partial x \tag{17}$$

shows that R plays exactly the same role as the Rayleigh number in thermal convection and measures the relative contributions of buoyancy forces to viscous plate driving forces. When R is large, buoyancy dominates and vigorous convection

Table 2. *Scaling relations for porosity driven convection*

(Primes denote dimensionless variables.)

variable	scale parameter	scaling relation
porosity	maximum porosity	$\phi = \phi_0 \phi'$
distance	melting depth	$(x, z) = d(x', z')$
		$\nabla = \nabla'/d$
velocity	plate velocity	$(\boldsymbol{v}, \boldsymbol{V}) = U_0(\boldsymbol{v}', \boldsymbol{V}')$
stream function		$\psi^s = dU_0 \psi^{s'}$

can occur. Unlike thermal convection however, it is horizontal variations in porosity, rather than temperature, that drive convection. Moreover, porosity is governed by a dispersive wave equation, while temperature is diffusive. When R is small, convection is negligible and mantle flow is dominated by the boundary conditions.

Figure 3a and b show the full steady-state solutions for the porosity, solid flow and melt flow for two end-member solutions in a geometry appropriate for ridges. These solutions are actually the end result of a long time dependent calculation that relaxes to steady state due to the free flux boundary condition at the top of the box. The possible time dependent effects of a freezing lid are discussed in §3b. In this solution, the dimensional melting rate is given by

$$\Gamma = \begin{cases} \rho_s F_{\max} W/d & W > 0, \\ 0 & W < 0 \quad \text{and on closed streamlines,} \end{cases} \tag{18}$$

which approximates melting by adiabatic decompression. $F_{\max}$ is the maximum degree of melting experienced on axis (here $F_{\max} = 0.25$), W is the vertical component of the matrix velocity and d is the depth of the melting region. Equation (18) states that any piece of fertile mantle melts at a rate proportional to the upwelling velocity, while material that moves sideways or down or along a closed flow line does not melt. Figure 3c, d shows the melting rate fields for these solutions.

When R is small (figure 3a) the matrix flow is essentially corner flow which can develop large non-hydrostatic pressure gradients that focus the flow of melt. The physics of melt focusing is discussed in detail in Spiegelman & McKenzie (1987) (see also Phipps Morgan 1987; Ribe 1988), who use an analytic constant porosity corner flow solution without melting. When melting is added, the porosity field is no longer constant, nevertheless, the basic behaviour of melt flow remains the same. As noted by earlier authors, this melt focusing provides a simple mechanism for extracting melt from a wide region at depth and extruding it at the surface in a narrow region. The principal geological problem with this solution, however, is that the actual melting region due to corner flow is quite wide and usually wider than the extraction zone, thus not all of the melt is extracted. Moreover, for reasonable estimates of melting depth and spreading rates, the viscosities required are large (i.e. 10^{20}–10^{21} Pa s). Such values have been considered 'unreasonable' although the rheology of two-phase melt-solid assemblages is still only poorly understood (Cooper & Kohlstedt 1984; Cooper & Kohlstedt 1986; Borch & Green 1990).

When viscosities are lower (more precisely for large values of R) then the focusing effect becomes negligible and internal sources of buoyancy become important. When the principal source of buoyancy in the melting region is due to lateral variations in melt content (Rabinowicz *et al.* 1984; Scott & Stevenson 1989; Buck & Su 1989) the effect is to produce an additional convective roll superposed on the plate driven

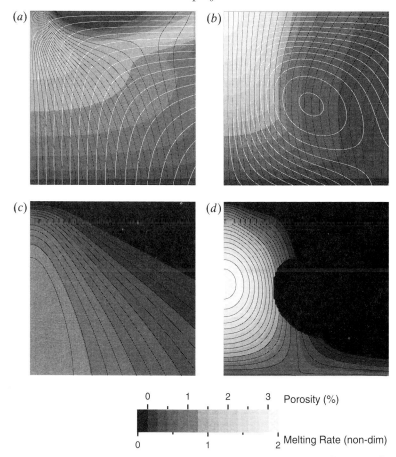

Figure 3. 2D solutions for the flow of melt and solid for a simple ridge spreading geometry demonstrating the effects of passive against active mantle flow. The top two parts show porosity, melt flow (black curves) and matrix flow (white curves). The bottom two figures show the dimensionless melting rate which is proportional to upwelling rate for 'fertile' upwelling mantle. $\Gamma = 0$ on closed streamlines or downwelling regions. Each part shows the right half of the melting region. The vertical scale is the melting depth (50–60 km) and there is no vertical exaggeration. R is related to the Raleigh number and measures the relative contribution of buoyancy forces to viscous forces. (*a*) Plate-dominated flow, $R = 0.0357$. When shear is dominated by boundary conditions, the forced shear of the viscous matrix can produce large pressure gradients that focus the flow of melt. (*b*) Buoyancy driven flow, $R = 56.28$. For lower viscosities or plate velocities, variations in melt content can drive additional small-scale convection which narrows the width of the upwelling region. (*c*) Melting rate field for plate driven shear, (*d*) and for buoyancy driven shear.

corner flow solution (figure 3*b*). This additional convection produces a narrow, fast upwelling zone beneath the ridge axis which narrows the region of melting. However, to produce a melting region with a width comparable to the neo-volcanic zone requires very large values of R which in turn requires very small matrix viscosities (Buck & Su 1989). For values of viscosity that are considered reasonable (10^{18}–10^{19} Pa s), the melting region is still approximately 20 km wide (figure 3*b*) and there is no additional melt focusing due to forced matrix shear. When the buoyancy is driven solely by depletion effects or temperature differences (Sotin & Parmentier 1989; Parmentier & Phipps Morgan 1990; Sparks & Parmentier 1990*b*, *a*; Cordery & Phipps Morgan 1992) the narrowing is much more subdued.

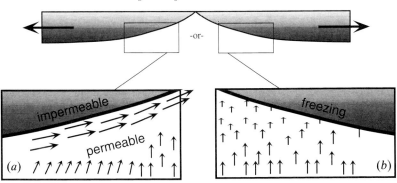

Figure 4. The two fundamental effects of freezing on the flow of melt near the top of the partly molten region. (*a*) Melt channelling: because the cooling plates form a natural sloping impermeable barrier to melt flow, channels may arise that transport melt to the ridge axis. (*b*) Simple freezing: as melt approaches the freezing boundary, it will be converted to solid which is transported away by the spreading plates.

The point of figure 3 is that both 'passive' and 'active' flow models are really end-member solutions of exactly the same equations. Both rely on the ability of the matrix to shear, the only difference between these two solutions is which of the two sources of shear dominate. If plate driving forces are sufficiently large, melt focusing may be sufficient to produce the narrow neo-volcanic zone. Alternatively, if the buoyancy forces are significant, then narrowed upwelling may play a part. It is not possible, however, to have both effects simultaneously, and neither process, by itself, appears to be sufficient for geologically reasonable values of R. Both processes, however, may be assisted by an additional mechanism for focusing that relies on compressible matrix deformation.

(b) *Compressible flow: freezing induced melt channels*

The principal result of §2*c* is that obstructions in the melt flux will shed solitary waves. At mid-ocean ridges one of the principal obstructions to the flow of melt is the presence of a frozen, impermeable lid produced by the spreading and cooling plates. This section will show that the solitary waves that develop due to freezing can form channels that guide the melt to the ridge axis.

One of the oldest suggestions for the lateral flow of melt at ridges is that melt percolates under gravity until it encounters the impermeable lid, develops some form of melt channel along the base of the sloping impermeable region and flows to the ridge axis. Figure 4*a* shows this channelling behaviour schematically. More recently, Sparks & Parmentier (1991) have begun to quantify this process and use results from a series of steady state 1D melt segregation problems to suggest that such a high porosity channel should exist. While this mechanism is appealing, the actual nature of the impermeable boundary is somewhat problematic. The principal problem is that the reason this boundary is impermeable is that it is simply too cold to allow the presence of any partial melt (i.e. it is a freezing boundary). Therefore, an equally plausible solution is that, as the melt approaches this boundary, it is converted to solid and is carried off by the spreading plates (figure 4*b*). It is not obvious which of these two effects of freezing actually controls the flow of melt near the top of the partly molten zone.

Rather than attempting to solve the full flow of melt and solid and its attendant temperature field beneath a ridge, figure 4 suggests that it may be sufficient to

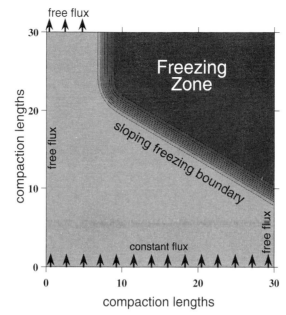

Figure 5. Geometry and boundary conditions used to investigate melt migration near a sloping freezing boundary. Freezing occurs in the darker region marked freezing zone. Contours and shading show values of the imposed freezing rate field $\Gamma(\boldsymbol{x})$. The freezing length, $\delta_{\mathrm{f}} = 1/\Gamma_0$ is approximately the length scale (in compaction lengths) over which a uniform melt flux would freeze completely. There is no matrix shear. Boundary conditions are constant melt flux into the base, free flux out the top and sides. The problem is non-dimensional with length scaled to the compaction length δ, porosity scaled to the uniform background porosity ϕ_0, and velocities to the background melt velocity w_0. Typical values for δ are 100–1000 m, for ϕ_0 are 1–10 % and for w_0 are 0.1–1 m a^{-1}.

understand the 2D flow of melt near a sloping freezing boundary. To that end, I have constructed a simple model problem which shows that both effects of freezing are actually possible and that again there is a single parameter that controls whether channels form.

Figure 5 illustrates the geometry and boundary conditions of the model. It consists of a two-dimensional region, 30 compaction lengths on a side. The freezing rate is imposed using the function

$$\Gamma = \tfrac{1}{4}\Gamma_0[1 - \tanh{(\boldsymbol{k}\cdot\boldsymbol{x})}][1 + \tanh{(k_1(x - x_1))}], \tag{19}$$

where Γ_0 is the maximum freezing rate in the interior of the freezing zone and $\boldsymbol{k}$ is a vector that is normal to the sloping boundary. The final hyperbolic tangent term is used to smoothly truncate the sloping freezing front at a distance x_1 from the left edge of the box. This truncation is used to avoid the interaction of the freezing zone with the boundaries of the region. While (19) appears somewhat complicated, the geometry is actually straight-forward and contours of Γ are shown in figure 5. For simplicity, this problem neglects any matrix shear deformation although, beneath a ridge, it is the large scale matrix flow that governs the thermal structure, and therefore the shape of the freezing region. Nevertheless, as long as the melt velocity is much greater than the matrix velocity, then the approximation of a fixed freezing region with no matrix shear is valid. Further experiments consider the effects of small amounts of matrix flow but do not change the basic results shown here.

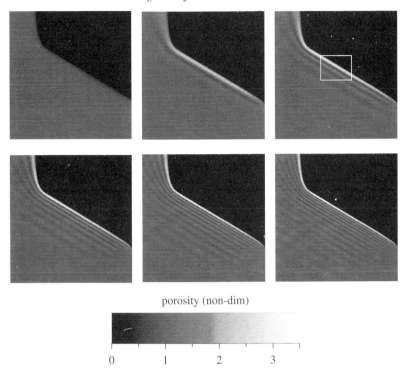

porosity (non-dim)

0   1   2   3

Figure 6. Evolution of porosity near a freezing zone with a rapid freezing rate ($\Gamma_0 = -2.0$) showing the development of high permeability melt channels near the freezing boundary. Each box is 30×30 compaction lengths on a side and shading shows porosity relative to the background porosity ($\phi = 1$) for dimensionless times $t = 0$, 10, 20 (top row), 40, 60, 80 (bottom row). A dimensionless time of 30 is the time required for a completely incompatible trace element to travel across the box at the background melt velocity. The initial condition is the steady-state solution given by the zero compaction length approximation where the melt can only percolate vertically and then freeze. At $t = 10$ the excess melt that cannot be accommodated by freezing begins to accumulate at the boundary in a growing channel ($\phi_{max} \approx 2.4$). With time the initial channel grows to a porosity approximately 3.5 times greater than the background. Rather than growing into a single channel, however, the growth of each channel initiates the formation of a new channel below it in the same manner as the dispersion of solitary waves. The white box at $t = 20$ is the region depicted in figure 7a.

Boundary conditions on the melt flux are a constant flux in to the bottom of the box, and the melt flux out of the top, right and left boundaries is free to adjust during the calculation. Detailed descriptions of numerical techniques and boundary conditions are given in Spiegelman (1993a, b). The initial condition for each run was the steady-state solution predicted by the zero compaction length approximation. In this approximation, the melt flux is driven solely by gravity and is exactly balanced by the freezing rate. Thus the initial condition for the porosity is given implicitly by the dimensionless flux balance

$$k_\phi(1 - \phi_0\phi) - (1 - \phi_0) = \int_0^z \frac{\rho_f \Gamma(x, z)}{\rho_s} \, dz \qquad (20)$$

for $\phi \geqslant 0$. Any variation from this initial condition during a calculation results from the term governing viscous resistance of the matrix to volume changes. This initial condition produces a freezing region of approximate 'freezing width' $\delta_f = 1/\Gamma_0$ over

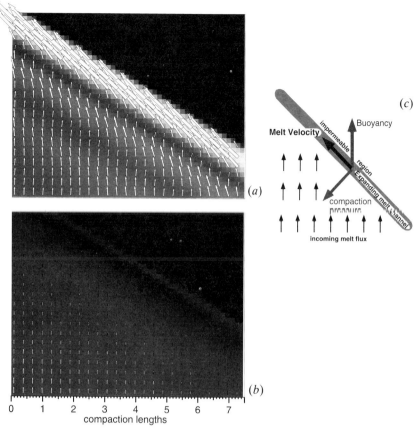

Figure 7. Close-up of the freezing boundary at $t = 20$ showing melt flux vectors for rapid and slow freezing rates. (*a*), (*b*) The 7.5×5.7 compaction length regions marked in figures 6 and 8. (*a*) Rapid freezing rate ($\Gamma_0 = -2$.): This figure shows that not only does the obstruction in flux cause porosity to increase near the freezing boundary, but also that the melt in these channels travels laterally. (*b*) When the freezing rate is slower ($\Gamma_0 = -0.1$), melt only percolates vertically. (*c*) Schematic diagram showing pressure gradients that make the melt flow laterally. When the melt flux is reduced by freezing over many compaction lengths, the compaction pressure is negligible and no channels form.

which the porosity decreases from the background porosity to impermeability. Note that the larger the freezing rate, the narrower the width of the freezing region. We now show that the larger the freezing rate, the more likely high permeability melt channels are to form.

(i) *Results*

Figure 6 shows the evolution of porosity for a rapid freezing rate and shows how the channels grow. When the freezing rate is large, the transition from the high permeability background to impermeability occurs over a distance comparable to the compaction length (here the width of the freezing layer is *ca.* 2δ) and the upwelling melt can 'see' the obstruction caused by the rapid change in permeability. Locally, the influx of melt is greater than can be consumed by freezing and the deformable matrix expands to accommodate the excess flux. This expansion leads initially to the growth of a high porosity channel near the freezing boundary ($t = 20$). Because the

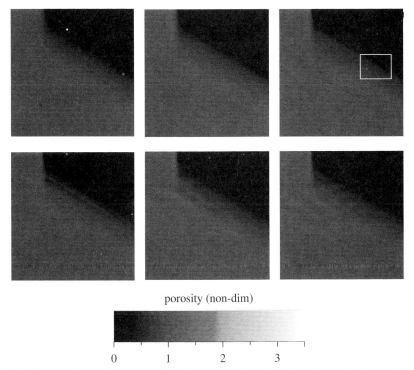

porosity (non-dim)

0 1 2 3

Figure 8. Evolution of porosity near a freezing zone with a slow freezing rate $\Gamma_0 = -0.1$ ($t = 0$, 10, 20 (top row), 40, 60, 80 (bottom row)). When freezing occurs over a region that is large compared to the compaction length, viscous pressure gradients are negligible and only weak channels form. With time, this solution develops a weak oscillation that is related to the dispersion of solitary waves (§2c), however, as a first approximation, this solution maintains steady state. The white box at $t = 20$ is the region depicted in figure 7b.

matrix is viscous, however, these volume changes generate additional pressure gradients that deflect the flow of melt away from the vertical and along the boundary. Figure 7a shows a close-up of melt flow in the channel at $t = 20$. In this problem, the dimensional melt flux is approximately

$$\phi \boldsymbol{v} = -k_\phi [(\zeta + \tfrac{4}{3}\eta)\nabla\mathscr{C} - \Delta\rho g\boldsymbol{k}]. \tag{21}$$

The term $(\zeta + \tfrac{4}{3}\eta)\mathscr{C}$ is the 'compaction pressure', the pressure induced by volume changes of the viscous matrix. The melt is driven vertically by buoyancy and away from the expanding channel by gradients of the compaction pressure which is always normal to the interface (see figure 7c). The net melt flux is therefore at an angle to the vertical. These compaction pressure gradients, however, only become significant when the melt flux varies over the compaction length. When freezing is distributed over many compaction lengths (figures 7b, 8), the compaction pressure is negligible. In this case, melt percolates vertically due to buoyancy alone and freezes into place without developing significant melt channels. Figure 8 shows the time evolution of porosity for gradual freezing and is steady state to a first approximation.

This problem shows that if the freezing zone is sufficiently narrow, high permeability melt channels can form to transport melt along the boundary. When channelling occurs, the flow is strongly time dependent. Rather than forming a single channel, a whole rank ordered set of channels forms, all oriented parallel to the

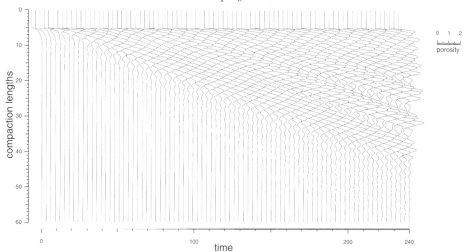

Figure 9. 1D dimensionless porosity profiles normal to a freezing boundary with a moderate freezing rate showing the development of a large dispersive train of melt channels. This calculation includes some flow of solid through the freezing region. The ratio of the solid velocity to the melt velocity is $W_0/w_0 = 0.01$, and freezing occurs over five compaction lengths from $z = 5$–10. The formation of these channels is similar to that for the dispersive solitary wave trains, however, these channels become absorbed in the freezing region. The channels form spontaneously without any time-dependent forcing. This problem appears to have no steady-state solution.

boundary. This behaviour is precisely the same phenomenon seen in the dispersion of solitary waves (§2c). In fact these channels are simply another manifestation of the solitary waves, however, they are pinned in place and take their overall form from the freezing boundary. Additional experiments suggest that, in addition to the two principally steady-state solutions shown here, there is also a strongly time-dependent régime where moderate amplitude channels can form but are transient. Figure 9 shows the detailed evolution of porosity taken in a 1D profile normal to the freezing boundary. The solution shown in figure 9 allows the matrix to advect porosity into the freezing region, which may affect the time dependence of the channels. In the time dependent régime, the initiation of channels can propagate far into the interior possibly producing an overall anisotropy in the large scale permeability that aligns itself with the freezing boundary. Lateral flow caused by any additional channels far from the freezing boundary, however, does not have to compete with freezing.

This simple model suggests that the existence or otherwise of channels in the Earth is controlled primarily by the ratio of the compaction length to the freezing length. Calculations based on the 1D steady-state thermal structure of an upwelling column (Sparks & Parmentier 1991; Spiegelman 1991) suggest that channelling may be an important mechanism for at least some lateral flow. However, more work needs to be done to consider the effects of channelling for larger-scale 2D and 3D problems. It should be stressed that this mechanism is completely different from the mechanisms that rely on incompressible shear (§3a) and can work in conjunction with both passive and active matrix flows. Compressible matrix deformation and the growth of channels are small-scale boundary layer phenomena that only depend on the local structure of the freezing zone. The structure of the freezing zone, however, depends on the large scale 2D and 3D flow and thermal structure of the mantle. This extreme variation in scales of processes is characteristic of the governing equations and makes accurate resolution of both compressible and incompressible effects numerically difficult. Nevertheless, if the local parameters that control channelling can be derived for any given solution for large-scale mantle flow, it becomes straightforward to discover whether melt channels influence lateral melt migration.

4. Summary

The purpose of this paper is to illustrate and clarify the wide range of behaviour inherent in the equations governing flow in viscously deformable porous media. It should be stressed this behaviour arises because the matrix is permeable and can deform viscously. Provided that the melt is interconnected everywhere, these equations form a macroscopic description of magma migration and make no assumptions about the microscopic distribution of melt. Clearly, as the functional form of the permeability and rheology change, the quantitative solutions to the equations will also change.

It is the ability for the matrix to deform viscously that adds most of the interesting new behaviour to these equations. In particular, this paper demonstrates that the matrix can deform in two fundamentally different ways. Because the matrix can expand and compact, variations in melt flux can actually propagate through the matrix as porosity waves. These waves are an essential feature of the governing equations and will form spontaneously from any region where more melt is produced or injected than can be readily extracted. Compressible matrix deformation also controls the flow of melt around obstacles such as the impermeable top of the partly molten zone. From the simple solutions shown here, it is clear that the compaction length is the only intrinsic length scale governing compressible matrix deformation. While the compaction length can be quite small it cannot in general be neglected. Rather, these problems show that a small compaction length implies that viscous effects will become important in narrow boundary layers. However, because of the dispersive behaviour of these equations, initially localized boundary layers can propagate away from the region of initial disturbance. A surprising but potentially important result is that *information* about obstructions in the melt flux can propagate backwards while the melt can only propagate in a forward direction. Eventually the entire partly molten region can become affected. Because of this behaviour, it is possible to construct simple initial conditions that never relax to steady state (see figures 2 and 9).

The second mode of deformation is incompressible matrix shear which controls porosity driven convection. At ridges the vigour of convection is determined by the balance of shear driven by boundary conditions and internal buoyancy forces. While porosity driven convection is superficially similar to thermal convection, it should be stressed that porosity does not behave in the same manner as temperature. Therefore, the stability of two-phase regions to small perturbations may be quite different from that expected in thermal convection. It should also be noted that incompressible matrix deformation is sensitive to the largest length scales in the problem, namely the size of the partly molten region. However, compressible flow is sensitive to the smallest length scale. Because of this disparity in length scales, accurate large-scale numerical solutions for the flow of both melt and solid become computationally expensive. For this reason, most numerical solutions for incompressible two-phase flow either neglect viscous compaction effects or neglect the flow of melt altogether. Nevertheless, the interaction between incompressible and compressible matrix deformation needs to be better understood. For example, large-scale solid flow controls the structure of the freezing boundary and therefore of the behaviour of the melt channels. Because the melt channels are dispersive, however, it is possible for compressible effects to grow to scales that might influence porosity driven convection. Because both incompressible and compressible mechanisms can contribute to melt focusing, it is important to understand the behaviour of the full set of equations.

Given the wide range of behaviour inherent in the governing equations, some of the geological implications of these results are clear. The melt extraction process should be inherently episodic in space and time. This episodicity should manifest itself in the volume, geometry and timing of eruptions. Given accurate forward solutions, it is straightforward to calculate the effects of geologically reasonable two-phase flows on geophysical and geochemical observables. For example, if we understood the coupling between melt content and seismic velocity, we could determine how well partly molten zones can be imaged. Present work is considering the effect of melt segregation on trace element chemistry and suggests that the spatial and temporal variations in chemistry may be a good indicator of the geodynamic process occurring at depth (Spiegelman 1992).

Many thanks to Dan McKenzie and Tim Elliot for useful discussions. This work was supported by NSF Grants OCE90-12572 and OCE91-14959 and is Lamont-Doherty Geological Observatory Contribution no. 4959.

References

Barcilon, V. & Lovera, O. 1989 Solitary waves in magma dynamics. *J. Fluid. Mech.* **204**, 121–133.

Barcilon, V. & Richter, F. M. 1986 Non-linear waves in compacting media. *J. Fluid Mech.* **164**, 429–448.

Bear, J. 1988 *Dynamics of fluids in porous media*. New York: Dover.

Borch, R. S. & Green, H. W. 1990 Experimental investigation of the rheology and structure of partially molten lherzolite deformed under upper mantle pressures and temperatures. *Eos, Wash.* **71**, 629.

Buck, W. R. & Su, W. 1989 Focused mantle upwelling below mid-ocean ridges due to feedback between viscosity and melting. *Geophys. Res. Lett.* **16**, 641–644.

Burnett, M., Caress, D. & Orcutt, J. 1989 Tomographic image of the magma chamber at 12° 50′ N on the East Pacific Rise. *Nature, Lond.* **339**, 206–208.

Caress, D. W., Burnett, M. S. & Orcutt, J. A. 1992 Tomographic image of the axial low velocity zone at 12° 50′ N on the East Pacific Rise. *J. geophys. Res.* **97** (B6), 9243–9264.

Cheadle, M. 1989 Properties of texturally equilibrated two-phase aggregates. Ph.D., University of Cambridge.

Cooper, R. & Kohlstedt, D. 1984 Solution-precipitation enhanced diffusional creep of partially molten olivine-basalt aggregates during hot pressing. *Tectonophys.* **107**, 207–233.

Cooper, R. & Kohlstedt, D. 1986 Rheology and structure of olivine-basalt partial melts. *J. geophys. Res.* **91**, 9315–9323.

Cordery, M. & Phipps Morgan, J. 1992 Melting and mantle flow beneath a mid-ocean spreading center. *Earth planet. Sci. Lett.* **111**, 493–516.

Daly, S. F. & Richter, F. M. 1989 Dynamical instabilities of partially molten zones: solitary waves vs. Rayleigh Taylor plumes. *Eos, Wash.* **70**, 499.

Detrick, R., Buhl, P., Vera, E., Mutter, J., Orcutt, J., Madsen, J. & Brocher, T. 1987 Multi-channel seismic imaging of a crustal magma chamber along the East Pacific Rise. *Nature, Lond.* **326**, 35–41.

Drew, D. 1971 Average field equations for two-phase media. *Stud. appl. Math.* **50**, 133–166.

Drew, D. 1983 Mathematical modeling of two-phase flow. *An. Rev. Fluid Mech.* **15**, 261–291.

Dullien, F. 1979 *Porous media fluid transport and pore structure.* New York: Academic Press.

Fowler, A. 1985 A mathematical model of magma transport in the asthenospere. *Geophys. Astrophys. Fluid Dyn.* **33**, 63–96.

MacDonald, K. 1982 Mid-ocean ridges: fine scale tectonic, volcanic, and hydrothermal processes within the plate boundary zone. *A. Rev. Earth planet. Sci.* **10**, 155–190.

McKenzie, D. 1984 The generation and compaction of partially molten rock. *J. Petrol.* **25**, 713–765.

Parmentier, E. M. & Phipps Morgan, J. 1990 The spreading rate dependence of three-dimensional spreading center structure. *Nature, Lond.* **348**, 325–328.

Phipps Morgan, J. 1987 Melt migration beneath mid-ocean spreading centers. *Geophys. Res. Lett.* **14**, 1238–1241.

Rabinowicz, M., Nicola, A. & Vigneresse, J. 1984 A rolling mill effect in the asthenosphere beneath oceanic spreading centers. *Earth planet. Sci. Lett.* **67**, 97–108.

Ribe, N. 1985 The deformation and compaction of partially molten zones. *Geophys. Jl R. astr. Soc.* **83**, 137–152.

Ribe, N. 1988 On the dynamics of mid-ocean ridges. *J. geophys. Res.* **93**, 429–436.

Ribe, N. & Smooke, M. 1987 A stagnation point flow model for melt extraction from a mantle plume. *J. geophys. Res.* **92**, 6437–6443.

Richter, R. M. & McKenzie, D. 1984 Dynamical models for melt segregation from a deformable matrix. *J. Geol.* **92**, 729–740.

Scheidegger, A. E. 1974 *The physics of flow through porous media.* University of Toronto Press.

Scott, D. 1988 The competition between percolation and circulation in deformable porous medium. *J. geophys. Res.* **93**, 6451–6462.

Scott, D. & Stevenson, D. 1984 Magma solitons. *Geophys. Res. Lett.* **11**, 1161–1164.

Scott, D. & Stevenson, D. 1986 Magma ascent by porous flow. *J. geophys. Res.* **91**, 9283–9296.

Scott, D. & Stevenson, D. 1989 A self-consistent model of melting, magma migration, and buoyancy-driven circulation beneath mid-ocean ridges. *J. geophys. Res.* **94**, 2973–2988.

Scott, D., Stevenson, D. & Whitehead, J. 1986 Observations of solitary waves in a deformable pipe. *Nature, Lond.* **319**, 759–761.

Sotin, C. & Parmentier, E. M. 1989 Dynamic consequences of compositional and thermal density stratification beneath spreading centers. *Geophys. Res. Lett.* **16**, 835–838.

Sparks, D. W. & Parmentier, E. M. 1990*a* 3-D flow beneath spreading centers due to compositional and thermal buoyancy. *Eos, Wash.* **71**, 1637.

Sparks, D. W. & Parmentier, E. M. 1990*b* Buoyant flow beneath ridge-transform systems and along-axis variations in gravity and crustal thickness. *Eos, Wash.* **71**, 627.

Sparks, D. W. & Parmentier, E. M. 1991 Melt extraction from the mantle beneath spreading centers. *Earth planet. Sci. Lett.* **105**, 368–377.

Spiegelman, M. 1991 2-D or not 2-D: understanding melt migration near a sloping, freezing boundary. *Eos, Wash.* **72**, 265.

Spiegelman, M. 1992 Passive vs. active flow? Only the tracers know… *Eos, Wash.* **73**, 290.

Spiegelman, M. 1993*a* Flow in deformable porous media. Part 1. Simple analysis. *J. Fluid Mech.* **247**, 17–38.

Spiegelman, M. 1993*b* Flow in deformable porous media. Part 2. Numerical analysis – the relationship between shock waves and solitary waves. *J. Fluid Mech.* **247**, 39–63.

Spiegelman, M. & McKenzie, D. 1987 Simple 2-D models for melt extraction at mid-ocean ridges and island arcs. *Earth planet. Sci. Lett.* **83**, 137–152.

Toomey, D. R., Purdy, G. M., Solomon, S. C. & Wilcox, W. S. D. 1990 The three-dimensional seismic velocity structure of the East Pacific Rise near latitude 9° 30′ N. *Nature, Lond.* **347**, 639–645.

Von Bargen, N. & Waff, H. S. 1986 Permeabilities, interfacial areas and curvatures of partially molten systems: results of numerical computations of equilibrium microstructures. *J. geophys. Res.* **91**, 9261–9276.

A laboratory study of melt migration

By M. J. Daines and D. L. Kohlstedt

*Department of Geology and Geophysics, University of Minnesota, Pillsbury Hall,
Minneapolis, Minnesota 55455, U.S.A.*

Melt migration experiments have been carried out to investigate the dependence of
permeability on melt fraction and on orthopyroxene content for partly molten
aggregates of olivine plus basalt and olivine-orthopyroxene plus basalt. Melt
migration couples, formed between discs of olivine or olivine with 20 vol %
orthopyroxene plus *ca.* 8 vol % basalt and discs with either *ca.* 2 vol % or 0 vol %
basalt, were heated at 1250 °C and 300 MPa for 6–32 h. The resulting melt migration
profiles, which developed in response to capillary forces, were analysed in terms of
coupled differential equations describing melt migration via porous flow through a
deformable matrix (compaction theory). For the experiments in which melt was
initially present in both the source and the sink, the melt migration profiles could be
fit equally well with a permeability proportional to the melt fraction to the first,
second or third power. For the experiments in which the sink initially was melt-free,
a best fit to the melt migration profiles could be obtained with a permeability that
is linearly proportional to melt fraction. The melt migration profiles for samples in
which the sink disc contained olivine plus 20 vol % orthopyroxene were essentially
identical to those for samples in which the sink contained only olivine, even though
at least some of the triple junctions in the former samples were not wetted by the
melt phase.

1. Introduction

The rate at which melt percolates through partly molten regions of the mantle affects
both the chemical composition and the physical properties of the host rock (cf.
Richter & McKenzie 1984; McKenzie 1989; Kohlstedt 1992). In a rock undergoing
pressure-release melting beneath a mid-ocean ridge, for example, the melt fraction
will increase with increasing degree of melting, if the permeability is low. Such an
increase in melt fraction could result in a marked decrease in viscosity (van der Molen
& Paterson 1979). In contrast, if the permeability is relatively high such that the
melt fraction remains low (e.g. below *ca.* 5 vol %), then the melt will have very little
effect on the rheological behaviour of the partly molten rock (Cooper & Kohlstedt
1986; Cooper *et al.* 1989).

The ability to establish an interconnected melt network in partly molten rocks is
a prerequisite for melt transport by porous flow and depends on the grain-scale
distribution of melt. Melt distribution can be characterized by the dihedral angle,
which is a function of the relative values of the solid–solid and solid–melt interfacial
energies in the rock. In mantle analogues composed of olivine plus silicate melt,
reported dihedral angles range from 20 to *ca.* 50° (cf. Waff & Bulau 1979, 1982;

Toramaru & Fujii 1986), implying that silicate melt in an olivine matrix is completely interconnected with melt confined to three and four grain junctions (Smith 1964; Beere 1975; Wray 1976; Bulau *et al.* 1979). This interpretation of these values for the dihedral angle has been experimentally confirmed (Vaughan *et al.* 1982; Daines & Richter 1988; Kohlstedt 1990). The upper mantle, although composed predominantly of olivine, also contains other crystalline phases, such as orthopyroxene, which may affect the establishment of an interconnected melt network. Microstructural studies of partly molten peridotites (Toramaru & Fujii 1986; Fujii *et al.* 1986; von Bargen & Waff 1988) suggest that, depending on water content, dihedral angles of melt in contact with orthopyroxene could be significantly higher than those for basalt in contact with olivine, potentially resulting in imperfect melt connectivity and a reduction in permeability (Toramaru & Fujii 1986; Nakano & Fujii 1989).

In an earlier series of experiments designed to investigate the kinetics of melt migration in aggregates of olivine plus a silicate melt, melt was induced to flow from a melt-rich source to a melt-free sink in response to the capillary forces arising from the gradient in melt fraction (Watson 1982; Riley & Kohlstedt 1990; Riley & Kohlstedt 1992). To determine a permeability as a function of melt fraction, the resulting melt migration profiles were analysed in terms of the compaction theory that was developed to describe porous flow through a deformable matrix (McKenzie 1984), but modified to replace buoyancy with a capillary driving force (Riley *et al.* 1990; Riley & Kohlstedt 1991). One striking and unanticipated result of these studies was a linear dependence of permeability, K, on melt fraction, ϕ (i.e. $K \propto \phi^1$). In contrast, permeability models yield power-law relations between permeability and melt fraction of the form

$$K(\phi) = (d^2/b)\,\phi^n \tag{1}$$

with $2 \leqslant n \leqslant 3$ (Turcotte & Schubert 1982, pp. 383–384; McKenzie 1984; Dullien 1992, ch. 3). In equation (1), d is the grain size and b is a geometrical constant. In particular, Cheadle (reported in McKenzie (1989)) analysed the permeability of a melt–solid system for which the pore geometry is texturally equilibrated (i.e. determined by the relative values of the solid–solid and solid–melt interfacial energies (cf. von Bargen & Waff 1986)); this analysis yielded $n = 2$ for $\phi < 0.03$ and $n = 3$ for $\phi > 0.03$.

For melt migration experiments in which the sink for melt initially is free of melt, one complication in applying a continuum formalism such as compaction theory is associated with the form of the capillary driving force

$$\frac{\partial}{\partial z}\left(-\frac{\partial G_v}{\partial \phi}\right) = \frac{\gamma_{sm}}{\sqrt{3}d}\frac{H(\theta)}{\phi^{\frac{3}{2}}}\frac{1}{}\frac{\partial \phi}{\partial z}, \tag{2}$$

where G_v is the free energy per unit volume associated with the presence of solid–solid and solid–melt interfaces, γ_{sm} is the solid melt interfacial energy, $H(\theta)$ is a function of the dihedral angle θ (cf. Riley & Kohlstedt 1991), and z is the spatial coordinate in the direction of melt flow. Based on equation (2), the driving force for melt migration approaches infinity as the melt fraction in the sink goes to zero.

Hence in the present study, two series of experiments were undertaken, one to eliminate the complication associated with a melt-free sink and the other to explore the effect of enstatite on the kinetics of melt migration. In the first set of experiments, melt migration couples were prepared in which not only the source, but

Table 1. *Composition of melt and crystalline phases used in melt migration experiments*

oxide component	MORB[a] (wt %)	Alkali basalt[b] (wt %)	Balsam Gap dunite[a] (wt %)	Bamble enstatite (wt %)	San Carlos olivine (wt %)
SiO_2	50.02	46.5	40.5	56.2	39.66
MgO	9.09	8.7	51.1	32.4	51.1
FeO	9.24	12.4	7.51	9.7	9.1
CaO	11.51	9.9	0.01	0.2	0.14
Al_2O_3	15.63	14.9	—	0.07	0.07
Na_2O	3.12	3.0	—	0	0.01
K_2O	0.27	0.97	—	0.07	—
TiO_2	1.55	0.97	—	0.07	—
P_2O_5	0.21	2.4	—	0.05	—
Cr_2O_3	0.05	0.3	—	—	—
NiO	—	—	0.31	—	—

[a] Cooper & Kohlstedt (1984).
[b] Helz (1973).

also the sink contained melt; the melt fraction in the sink was about one-fifth that in the source. In the second set of experiments, the crystalline matrix was composed of olivine plus orthopyroxene. The resulting melt migration profiles from both types of experiments were analysed in terms of compaction theory to determine the dependence of permeability on melt fraction.

2. Experimental procedure

Melt migration couples for the two different types of experiments were formed by pressing together two polycrystalline discs, one a source for melt and the other a sink for melt. In the first set of experiments, both source and sink discs were composed of olivine plus alkali basalt to determine the dependence of permeability on melt fraction. Source discs contained *ca.* 8 vol % melt and the sink discs *ca.* 2 vol % melt. In the second set of experiments, source discs were composed of dunite with *ca.* 20 vol % enstatite plus basalt, whereas sink discs were melt-free containing either olivine or dunite plus *ca.* 20 vol % enstatite to investigate the effect of orthopyroxene on the kinetics of melt migration. All discs were separately fabricated by isostatically hot-pressing mechanical mixtures of powders of the appropriate composition. Starting powders with a particle size of less than 10 μm were obtained by mechanically grinding San Carlos olivine, Balsam Gap dunite, Bamble enstatite, a mid-ocean ridge basalt (MORB) and an alkali basalt. Because the powers used in our experiments were dried at only 150 °C, compaction and melt migration most likely occurred under 'wet' conditions. Compositions of starting materials are listed in table 1.

To synthesize texturally and chemically equilibrated source and sink discs with a density of 99 % of the theoretical value, each disc was held at *ca.* 1250 °C for at least 3 h at 300 MPa confining pressure in a gas-medium high-pressure vessel. The discs were jacketed in Ni to control the oxygen partial pressure within the olivine stability field at the Ni–NiO phase boundary. For the melt migration experiments, each melt migration couple (encapsulated in Ni) was held at *ca.* 1250 °C, 300 MPa for between

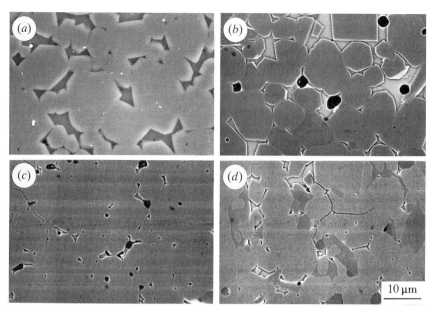

Figure 1. SEM micrographs of typical microstructures in melt migration couples. Glass appears as darker grey regions, although the central portion of larger glass pockets in back-scattered electron images (b), (c) and (d) are very light. Olivine is light grey; orthopyroxene is an intermediate grey. The darkest regions are pores or small cracks in the sample. (a) Secondary electron image of the source region in the 32 h olivine–alkali basalt melt migration experiment. (b) Back-scattered electron image of the source region in the ol–opx–MORB experiment. (c) Back-scattered image of the sink that initially contained only olivine in the ol–opx–MORB experiment. (d) Back-scattered electron image of the sink that initially contained olivine and orthopyroxene in the ol–opx–MORB experiments.

6 and 32 h. The couples, each *ca.* 7 mm in diameter and *ca.* 2 mm in length, were then cut normal to the interface between the source and the sink discs and polished to a 0.05 μm finish.

Median dihedral angles were determined by a statistical analysis of 250 measurements of triple junctions on scanning electron microscope (SEM) micrographs. Grain diameters were measured on an optical microscope as well as on SEM micrographs; at least 100 measurements were made to determine mean grain sizes for source and sink discs. Melt migration profiles were subsequently obtained with an image analysis system associated with the SEM.

Each melt migration profile was analysed in terms of compaction theory for porous flow through a deformable matrix (McKenzie 1984), modified by replacing the buoyancy driving force with a capillary driving force (Riley & Kohlstedt 1991). A comparison of the experimentally determined melt migration profiles with those generated from simulations based on compaction theory yielded values for the compaction length, δ_c, and the characteristic time, τ_0. These fitting parameters are related to the permeability at the starting melt fraction in the source, K_0, the viscosity of the partly molten rock, η_{pm}, and the viscosity of the melt, μ, as well as d, $H(\theta)$ and γ_{sm}:

$$\delta_c = (K_0 \eta_{pm}/\mu)^{\frac{1}{2}}, \tag{3}$$

$$\tau_0 = \sqrt{3}\, d\eta_{pm}/\gamma_{sm} H(\theta). \tag{4}$$

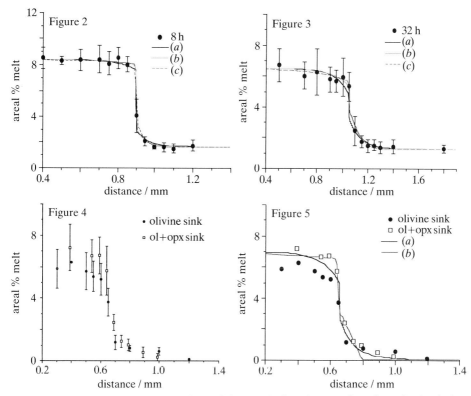

Figure 2. Plot of the melt fraction profile and final melt distributions from best-fit simulations for $n = 1$, 2 and 3 for the 8 h olivine plus alkali basalt experiment in which the sink initially contained some melt. (a) $n = 1$, $\delta_c = 90\ \mu m$; (b) $n = 2$, $\delta_c = 150\ \mu m$; (c) $n = 3$, $\delta_c = 200\ \mu m$.

Figure 3. Plot of the melt fraction profile and final melt distributions from best-fit simulations for $n = 1$, 2 and 3 for the 32 h olivine plus alkali basalt experiment in which the sink originally contained some melt. (a)–(c) As for figure 2.

Figure 4. Plot of the melt fraction profiles for olivine–orthopyroxene–MORB experiments in which one sink contained only olivine and the other contained only olivine plus orthopyroxene (i.e. the sinks were initially melt-free).

Figure 5. Plot of the melt fraction profiles for olivine–orthopyroxene–MORB experiments and final melt distributions from best-fit simulations for $n = 1$ and 3. (a) $n = 1$, $\delta_c = 150\ \mu m$; (b) $n = 3$, $\delta_c = 500\ \mu m$.

3. Results

Typical microstructures of samples are shown in figure 1. Both large glass pockets and small triangular triple junctions are present in all samples; the frequency and size of the large glass pockets decreases as glass fraction decreases. Olivine (ol) and orthopyroxene (opx) grains in contact with larger glass pockets tend to be faceted; as the size of the glass pockets decreases, these interfaces become more smoothly curved. All of the triple junctions that were examined in samples containing only olivine appear to contain some glass. In samples containing orthopyroxene, however, some triple junctions in which orthopyroxene is present appear to be glass-free. It is important to note that glass pockets smaller than $0.01\ \mu m$ are not detectable because of the resolution of the imaging methods used. Similar microstructures have also been described by von Bargen & Waff (1988). Grain size increased in all

Table 2. *Parameters obtained from melt migration experiments at* 1250 °C *with a sink of olivine plus* 2 *vol* % *basalt*

$\dfrac{t}{10^4 \text{ s}}$	n	$\dfrac{\tau_0}{10^6 \text{ s}}$	$\dfrac{\delta_c}{\mu\text{m}}$	$\dfrac{\eta_{pm}}{(10^{10} \text{ Pa s})}$	$\dfrac{K_0}{(10^{-18} \text{ m}^2)}$	$\dfrac{d_0}{\mu\text{m}}$	$\dfrac{d_f^{src}}{\mu\text{m}}$	$\dfrac{d_f^{snk}}{\mu\text{m}}$
2.88	1	11.52	90	9.6	0.84	9.0	10.5	10.5
	2		150		2.3			
	3		200		4.2			
11.52	1	11.52	90	6.3	1.3	9	13.5	16.5
	2		150		3.6			
	3		200		6.3			

Table 3. *Parameters obtained using* $n = 1$ *from melt migration experiments at* 1250 °C *with a sink of olivine or olivine plus* 20 *vol* % *orthopyroxene*

sink material	$\dfrac{t}{10^4 \text{ s}}$	$\dfrac{\tau_0}{10^6 \text{ s}}$	$\dfrac{\delta_c}{\mu\text{m}}$	$\dfrac{\eta_{pm}}{10^{10} \text{ Pa s}}$	$\dfrac{K_0}{10^{-18} \text{ m}^2}$	$\dfrac{d_0}{\mu\text{m}}$	$\dfrac{d_f^{src}}{\mu\text{m}}$	$\dfrac{d_f^{snk}}{\mu\text{m}}$
ol+opx	2.16	4.32	200	4.6	8.7	9.0	11.4	9.3
		2.88	150	3.0	7.5			
		2.16	100	2.2	4.5			
ol	2.16	8.64	200	7.7	5.2	9.0	13.5	13.5
		4.32	150	3.9	5.8			
		2.88	100	2.6	3.8			

samples during the melt migration experiments (see tables 2 and 3). The median dihedral angle for ol–ol–alkali basalt contacts was *ca.* 38°; for ol–ol–MORB, 27°; for ol–opx–MORB, 34°; for opx–opx–MORB, 41°. The median angle reported for opx–opx–MORB contacts is based on only 45 measurements due to the relative scarcity of these contacts.

Melt migration profiles, plotted as areal percent melt against distance in figures 2–5, illustrate that melt is transported from the source into the sink in all experiments. Within experimental uncertainties, the decrease in the amount of melt in the source is equal to the increase in the amount in the sink. Comparison of figure 2 with figure 3 demonstrates that the amount of melt that migrates from the source into the sink during an experiment increases with increasing time. The effect of the addition of orthopyroxene on the extent of melt migration is illustrated in figure 4.

Also presented in figures 2, 3 and 5 are the melt migration profiles obtained from compaction theory for values of $n = 1$, 2 and 3, where n is defined in equation (1) (i.e. $K \propto \phi^n$). For the experiments in which the sink initially contained some melt, the values of the parameters used to non-dimensionalize the compaction equations are summarized for each value of n in table 2. Values for the K_0 and η_{pm}, calculated from equations (3) and (4) using $\mu = 10$ Pa s (Ryan & Blevins 1987) and $\gamma_{sm} = 0.5$ J m^{-2} (Cooper & Kohlstedt 1982), are also listed. In tables 2 and 3, d_0, d_f^{src} and d_f^{snk} are the starting grain size in the melt migration couple, the final grain size in the source disc and the final grain size in the sink disc, respectively.

For the experiments in which the sink initially was melt-free, the values of the parameters used to non-dimensionalize the compaction equations are summarized in table 3. Because $n = 1$ clearly provides the best fit for these experiments (figure 5), only the values of the non-dimensionalization parameters obtained when $n = 1$ are reported. Again, values of $\mu = 10$ Pa s and $\gamma_{sm} = 0.5$ J m^{-2} were used to calculate K_0 and η_{pm} for these experiments.

4. Discussion

(a) *Effect of a melt-bearing sink*

Based on these experiments it is not possible to determine the dependence of permeability on melt fraction, except in those experiments in which the sink did not initially contain melt. As seen in figures 2 and 3, the experimental data can be fitted equally well with $n = 1$, 2 or 3. In part this lack of sensitivity in these experiments to the value of n could have been anticipated. Since $K \propto \phi^n$, the larger the value of ϕ, the smaller the order of magnitude change in permeability as n increases. In other words, as the melt fraction increases, the model and hence the experiment become insensitive to the difference in permeability that results from choosing different values of n. Consequently, it is only at very small melt fractions that the choice of n greatly affects the shape of the melt migration profile. In figures 2 and 3 it is apparent that even in the case of perfect sample homogeneity and errorless analysis, it would be difficult to determine a unique value of n. It is still possible, however, to extract information about the kinetics of melt migration from these experiments.

Because the scaling parameter τ_0 does not depend on the choice of n, the viscosity of the partially molten sample η_{pm} can be determined. The values obtained for the experiments run for 8 and 32 h, 9.6×10^{10} and 6.3×10^{10} Pa s, respectively, compare well to the viscosity of *ca.* 1×10^{11} Pa s for 'wet', partly molten olivine deforming by grain boundary diffusional creep at 1250 °C for a grain size of *ca.* 10 μm (Karato 1986; Cooper *et al.* 1989; Kohlstedt & Chopra 1993). The difference between the values for η_{pm} determined for the two experiments arises because, in fitting the data with the numerical simulations it was assumed that the time scaling parameter τ_0 should be constant for both experiments. Since $\eta_{pm} \propto \tau_0/d$ (equation (4)), the difference in the final mean grain size between the samples results in a small difference in η_{pm}, with the sample with the larger grain size having the lower viscosity. In fact, because for $\eta_{pm} \propto d^3$ in the grain boundary diffusion creep régime (Rutter 1976), it is likely that the matrix in the 32 h experiment, which had a larger final grain size, was more viscous than the matrix in the 8 h experiment. Even though this dependence of viscosity on grain size introduces some uncertainty into the determination of η_{pm}, it will not likely affect the value by more than a factor of 2.

Although the value of η_{pm} obtained is not dependent on the value of n, the value of K_0 increases by a factor of 5 as n is increased from 1 to 3. However, this variation is small compared to the discrepancy between the average value for K_0 obtained in our study, *ca.* 3×10^{-18} m^2, and that determined by Riley *et al.* (1990) and Riley & Kohlstedt (1991), *ca.* 2×10^{-15} m^2. One of the differences between the present experiments and the earlier experiments is the initial presence of melt in the sink disc. This difference, however, can not explain the discrepancy, because the values for K_0, *ca.* 4 to 8×10^{-18} m^2, obtained in the ol–opx–MORB experiments, in which the sink initially contained no melt, are also significantly smaller than those of Riley *et al.* (1990) and Riley & Kohlstedt (1991). The major uncertainty in calculating K_0 – and hence the likely source of the discrepancy – is in the values used for the viscosity of the melt phase μ. In the earlier experiments, the silicate melt was rich in potassium and aluminium; in the present experiments, the potassium content was almost negligible and the aluminium content was about a factor of two smaller. In both cases, values for μ were calculated from published compilations of melt viscosity (Ryan & Blevins 1987); consequently, uncertainties in these values could be large. Independent measurements of the viscosity of the melt phases are clearly needed.

The implications of the much smaller K_0 determined in the present experiments can be illustrated by calculating the length scale over which melt segregation via porous flow will be rapid enough to maintain the low resident melt fraction of 0.1% which was calculated by Riley *et al.* (1990). It was assumed in the original calculations that the melt production rate in a 50 km melting column under a mid-ocean ridge is equal to the melt extraction rate and that permeability is linearly dependent on melt fraction. Using the same melt production rate and requiring that the resident melt fraction be 0.1%, substitution of the permeability obtained in the present study for those of Riley *et al.* (1990) gives a length scale of *ca.* 20 m over which porous flow will occur rapidly enough to keep the melt fraction low. This length scale can be increased by either allowing the resident melt fraction to be higher or by decreasing the melt production rate. The implication of this smaller length scale over which porous flow is an effective mechanism of melt segregation is that there must be a transition from porous to channelized flow in the mantle on a comparable length scale.

(b) Effect of the addition of orthopyroxene

The addition of orthopyroxene does not have a substantial effect on the kinetics of melt migration; at most, there may be a very slight enhancement in the melt migration rate for samples containing orthopyroxene (figure 4). Microstructural evidence would suggest, however, that melt migration should actually be retarded in samples containing orthopyroxene due to the presence of some melt-free triple junctions (figure 1), which should increase the tortuosity of melt migration paths and consequently reduce the permeability. Toramaru & Fujii (1986) predicted that, in rocks in which olivine and orthopyroxene grain sizes are similar, complete melt connectivity is not obtained once the orthopyroxene content exceeds 25 vol%. In our experiments, the grain sizes of olivine and orthopyroxene are approximately equal, so that the presence of 20% orthopyroxene may not be sufficient to produce a measurable effect on the permeability. In addition, Toramaru & Fujii's analysis was based on a dihedral angle of 76° for opx–opx–melt in a dry peridotite; the dihedral median angle for opx–opx–melt contacts was *ca.* 41° (excluding melt-free triple junctions) in the present study for apparently wet samples. (Note that it has previously been observed (Fujii *et al.* 1986; von Bargen & Waff 1988) that the presence of water reduces the dihedral angle between orthopyroxene and basalt.)

The viscosities for the partly molten aggregates determined from the melt migration profiles for the olivine and the olivine plus orthopyroxene samples are the same within error. This result is consistent with limited data that are available for comparison of the rheology of olivine and olivine plus orthopyroxene aggregates (Hitchings & Paterson 1989).

Finally, it should be emphasized that the best fit of the coupled differential equations describing compaction theory to the experimental melt migration profiles for the experiments in which the sink initially was melt-free yields $n = 1$ (i.e. $K \propto \phi^1$). This result thus confirms that previously reported by Riley *et al.* (1990) and Riley & Kohlstedt (1991). Model calculations using compaction theory demonstrate that if our new sample design in which both the source and the sink initially contain melt is to be used to constrain the value of n, then the melt fraction in the sink must be less than about one-tenth that in the source. These experiments are presently in progress in our laboratory.

Support from the National Science Foundation through grant EAR-8916438 is gratefully acknowledged.

References

Beere, W. 1975 A unifying theory of the stability of penetrating liquid phases and sintering pores. *Acta metall.* **23**, 131–138.

Bulau, J. R., Waff, H. S. & Tyburczy, J. A. 1979 Mechanical and thermodynamic constraints on fluid distribution in partial melts. *J. geophys. Res.* **84**, 6102–6108.

Cooper, R. F. & Kohlstedt, D. L. 1982 Interfacial energies in the olivine-basalt system. In *Advances in Earth and planetary sciences* (ed. S. Akimoto & M. H. Manghnani), vol. 12 (*High Pressure Research in Geophysics*), pp. 217–228. Tokyo: Center for Academic Publications.

Cooper, R. F. & Kohlstedt, D. L. 1984 Sintering of olivine and olivine-basalt aggregates. *Phys. Chem. Minerals* **11**, 5–16.

Cooper, R. F. & Kohlstedt, D. L. 1986 Rheology and structure of olivine-basalt partial melts. *J. geophys. Res.* **91**, 9315–9323.

Cooper, R. F., Kohlstedt, D. L. & Chyung, K. 1989 Solution-precipitation enhanced creep in solid–liquid aggregates which display a non-zero dihedral angle. *Acta metall.* **37**, 1759–1771.

Daines, M. J. & Richter, F. M. 1988 An experimental method for directly determining the interconnectivity of melt in a partially molten system. *Geophys. Res. Lett.* **15**, 1459–1462.

Helz, R. T. 1973 Phase relations of basalts in their melting range at $P_{H_2O} = 5$ kb as a function of oxygen fugacity. *J. Petrol.* **14**, 249–302.

Hitchings, R. S., Paterson, M. S. & Bitmead, J. 1989 Effects of iron and magnetite additions in olivine-pyroxene rheology. *Phys. Earth planet. Interior* **55**, 277–291.

Karato, S.-I., Paterson, M. S. & FitzGerald, J. D. 1986 Rheology of synthetic olivine aggregates: Influence of grain size and water. *J. geophys. Res.* **91**, 8151–8176.

Kohlstedt, D. L. 1990 Chemical analysis of grain boundaries in an olivine–basalt aggregate using high-resolution, analytical electron microscopy. In *The brittle–ductile transition in rocks: the Heard volume* (ed. A. G. Duba, W. B. Durham, J. W. Handin & W. F. Wang), pp. 211–218. Washington: American Geophysical Union.

Kohlstedt, D. L. 1992 Structure, rheology and permeability of partially molten rocks at low melt fractions. In *Mantle flow and melt generation at mid-ocean ridges* (ed. J. P. Morgan), 1990 vol. (*RIDGE Summer Institute*). Washington: American Geophysical Union.

Kohlstedt, D. L. & Chopra, P. N. 1993 Influence of basaltic melt on the creep of polycrystalline olivine under hydrous conditions. In *Magmatic systems* (ed. M. P. Ryan). New York: Academic Press.

McKenzie, D. 1984 The generation and compaction of partially molten rock. *J. Petrol.* **25**, 713–765.

McKenzie, D. 1989 Some remarks on the movement of small melt fractions in the mantle. *Earth planet. Sci. Lett.* **95**, 53–72.

Nakano, T. & Fujii, N. 1989 The multiphase grain control percolation: its implication for a partially molten rock. *J. geophys. Res.* **94**, 15.653–15.661.

Ribe, N. M. 1987 Theory of melt segregation – a review. *J. Volcan. geotherm. Res.* **33**, 241–253.

Richter, F. M. & McKenzie, D. 1984 Dynamical models for melt segregation from a deformable matrix. *J. Geol.* **92**, 729–740.

Riley, G. N. Jr & Kohlstedt, D. L. 1990 An experimental study of melt migration in an olivine–melt system. In *Magma transport and storage* (ed. M. P. Ryan), pp. 77–86. New York: Wiley.

Riley, G. N. Jr & Kohlstedt, D. L. 1991 Kinetics of melt migration in upper mantle-type rocks. *Earth planet. Sci. Lett.* **105**, 500–521.

Riley, G. N. Jr, Kohlstedt, D. L. & Richter, F. M. 1990 Melt migration in a silicate liquid–olivine system: an experimental test of compaction theory. *Geophys. Res. Lett.* **17**, 2101–2104.

Riley, G. N. Jr & Kohlstedt, D. L. 1992 The influence of H_2O and CO_2 on melt migration in two silicate liquid-olivine systems. In *Fault mechanics and transport properties in rocks* (ed. B. Evans & T.-F. Wong) (*A Symposium in Honor of W. F. Brace*), pp. 281–293. New York: Academic Press.

Rutter, E. H. 1976 The kinetics of rock deformation by pressure solution. *Phil. Trans. R. Soc. Lond.* A **283**, 203–219.

Ryan, M. P. & Blevins, J. Y. K. 1987 *The viscosity of synthetic and natural silicate melts and glasses at high temperatures and* 1 bar (10^5 Pa) *pressure and at higher pressures.* (466 pages.) Denver: U.S. Geological Survey.

Smith, C. S. 1964 Some elementary principles of polycrystalline microstructure. *Metall. Rev.* **9**, 1–47.

Toramaru, A. & Fujii, N. 1986 Connectivity of melt phase in a partially molten peridotite. *J. geophys. Res.* **91**, 9239–9252.

van der Molen, I. & Paterson, M. S. 1979 Experimental deformation of partially melted granite. *Contrib. Mineral. Petrol.* **70**, 291–318.

Vaughan, P. J., Kohlstedt, D. L. & Waff, H. S. 1982 Distribution of the glass phase in hot-pressed, olivine–basalt aggregates: an electron microscopy study. *Contrib. Mineral. Petrol.* **81**, 253–261.

von Bargen, N. & Waff, H. S. 1986 Permeabilities, interfacial areas and curvatures of partially molten systems: results of numerical computations of equilibrium microstructures. *J. geophys. Res.* **91**, 9261–9276.

von Bargen, N. & Waff, H. S. 1988 Wetting of enstatite by basaltic melt at 1350 °C and 1.0- to 2.5-GPa pressure. *J. geophys. Res.* **93**, 1153–1158.

Waff, H. S. & Bulau, J. R. 1979 Equilibrium fluid distribution in an ultramafic partial melt under hydrostatic stress conditions. *J. geophys. Res.* **84**, 6109–6114.

Waff, H. S. & Bulau, J. R. 1982 Experimental determination of near-equilibrium textures in partially molten silicates at high pressures. In *Advances in Earth and planetary sciences* (ed. S. Akimoto & M. H. Manghnani), vol. 12 (*High Pressure Research in Geophysics*), pp. 229–236. Tokyo: Center for Academic Publications.

Watson, E. B. 1982 Melt infiltration and magma evolution. *J. Geol.* **10**, 236–240.

Wray, P. J. 1976 The geometry of two-phase aggregates in which the shape is determined by its dihedral angle. *Acta metall.* **24**, 125–135.

Thermal and petrological consequences of melt migration within mantle plumes

By Georges Ceuleneer, Marc Monnereau, Michel Rabinowicz
and Christine Rosemberg

Groupe de Recherche de Géodésie Spatiale, Centre National de la Recherche Scientifique UPR 234, 14, av. Ed. Belin, 31400 Toulouse, France

The high temperatures and high degrees of melting expected in the core of mantle plumes have virtually no expression in the eruption temperatures of hotspot lavas, nor in the composition of their glasses, which is restricted in the basaltic field. A solution to this paradox is looked for in the melt migration processes within the melting region of mantle plumes. Three dimensional convective calculations at Rayleigh number of 10^6 allow estimates of the possible temperature, melt fraction and stress fields within a plume. Two regions with different melt migration patterns can be distinguished. A lower zone ranging in depth from the base of the melting region (150 km) to around 80–100 km where the first melt fraction is redistributed in a sub-horizontal vein network and convects in response to the steep horizontal temperature gradient. This process is able to homogenize the temperature within the melting region very efficiently. The high (300 °C) temperature contrast between the centre of the plume and the surrounding mantle can be reduced to a few tens of degrees at the top of this zone. Fractional crystallization of high pressure phases will strongly modify the composition of the melt as it circulates toward the periphery of the melting region. A second upper zone, where the sub-vertical vein orientation will make possible rapid melt migration toward the surface, extends to the base of the lithosphere. Due to the buffering of the plume temperature around a value close to the mean upper mantle temperature, the degree of adiabatic melting within this upper zone will not greatly exceed that beneath normal spreading centres, even in the case of on-ridge hotspots. The lavas erupted at hotspots are likely to result from the mixing in various proportions of these low pressure melts (basalts) with the highly evolved liquids (possibly with kimberlitic to alkalic affinities) resulting from fractional crystallization of the high-pressure melt fractions produced at the base of the melting region. This scenario could account for the low eruption temperatures and Mg contents of hotspot lavas, in spite of a complex high pressure, and thus high temperature, history evidenced by some geochemical trends.

1. Introduction

A challenge for the coming years is to establish accurate relationships between the solid-state mantle convection and the chemistry of the melts erupted at the Earth surface which, for a given mantle composition, is related to parameters like the degree of melting, the depth of melt segregation and the amount of fractional crystallization (Albarède 1992; Langmuir *et al.* 1992). In that spirit, McKenzie & Bickle (1988) suggest that the mid-ocean ridge volcanism provides a largely random

Phil. Trans. R. Soc. Lond. A (1993) **342**, 53–64

sampling of the upper mantle far from any deep-seated upwelling and allows to estimate a mean mantle potential temperature. Their estimation (1280 °C) falls within the range of values deduced from a worldwide survey of mid-ocean ridge basalts chemistry (Klein & Langmuir 1987). Hotspots constitute another geodynamic setting where mantle partial melts are produced in abundance. Geophysical and isotopic data are largely consistent with a 'plume' origin for hotspot volcanism. Compared to environments of passive upwelling like the oceanic ridges, a complex thermal structure is expected in the mantle underlying hotspot volcanoes. Since Morgan's hypothesis relating hotspots to ascending convective currents, our understanding of hotspot dynamics has been largely conditioned by progress in the modelling of mantle convection. Thanks to the recent increase of computer power, it is now possible to perform high resolution three-dimensional (3D) convective calculations and to account for some geophysical observations related to hotspots (Rabinowicz *et al.* (1990) and references therein), and thus to check in what extent the thermal fields predicted by such convective models are consistent with the petrological composition of hotspot lavas. A characteristic shared by most models of mantle plumes is an important temperature contrast – a few hundred degrees, at least – between the core of the plumes and an average mantle geotherm, which is consistent with the occurrence of partial melts below thick lithospheric lids (White & McKenzie 1989). Surprisingly, these very high temperatures have no expression in the measured eruption temperatures of hotspot lavas: maximum values of 1190 and 1270 °C are reported for the 'picritic' (i.e. highly olivine phyric) flows of Hawaii and Iceland respectively (Jakobsson *et al.* 1978; Helz 1987). These should be compared to the eruption temperatures calculated for primitive mid-ocean ridge basalts (1140–1210 °C; Allan *et al.* 1987). The temperatures of more common hotspot tholeiitic flows overlap the field of MORB's eruption temperatures. In the present paper, we propose a possible solution to this paradox. Based on recent high-pressure melting experiments and on a 3D numerical model of convection constrained by geophysical data, we estimate the nature of melts which should be produced within a mantle plume. To account for some basic petrological characteristics of hotspot lavas, we discuss possible thermal effects of melt migration within the plume.

2. Plume model

Previous attempts to account for a 'plume' convective pattern within the mantle have shown the importance of rheological stratification. A low viscosity zone (LVZ) ranging in depth from the base of the lithosphere to about 150–200 km is required by many geophysical observations and is likely to be related to the softening of mantle peridotites as they approach, and eventually cross, their solidus. Beneath hotspot swells, a viscosity contrast of about 50 between the LVZ and the underlying mantle has been inferred from geoid and bathymetric data (Robinson *et al.* 1987; Ceuleneer *et al.* 1988). Three-dimensional calculations of convection in large rectangular boxes have shown that such a LVZ induces the formation of a hexagonal convective planform where ascending flows have the form of steady narrow plumes (Rabinowicz *et al.* 1990). This pattern results mainly from the asymmetry induced by the low viscosity zone between the upper and lower convective boundary layers. It satisfies some of the observed (or expected) properties of hotspot plumes: the extreme narrowness of the upwelling currents (of the order of several tens of kilometres at the base of the lithosphere), their location at the centre of more diffuse downwelling

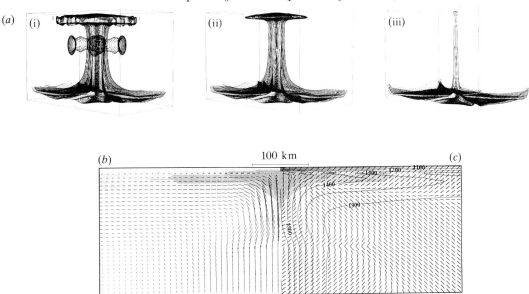

Figure 1. (*a*) Three isothermal surfaces in the area of ascending flow of the convective model: (i) 1290 °C, (ii) 1350 °C, (iii) 1500 °C. (*b*) Solidus (pale shading), cpx-out (intermediate shading) and opx-out (dark shading) boundaries according to Takahashi (1986) and Herzberg *et al.* (1990) for the upper third of the convective box in the area of ascending flow. The velocity field is shown on the left and the maximum compressive stress orientation is shown on the right.

currents, their stability through time spans of the order of hundreds of millions of years, their robustness relative to destabilizing influences like the drift of overlying plates, and their topographic and gravimetric signature.

The purpose of the present study is to compare a convective thermal field with the petrology of hotspot lavas, and we therefore need to reach realistic conditions for upper mantle convection. Accordingly, a 3D model with a Rayleigh number

$$Ra = g\alpha \Delta T d^3 / \kappa \nu \tag{1}$$

of 10^6 has been run. The gravity constant g is 10 m s^{-2}, the coefficient of thermal expansion α is 3.7×10^{-5} °C^{-1}, the upper mantle thickness d is 6.5×10^5 m and the thermal diffusivity κ is 8×10^{-7} m^2 s^{-1}. The kinematic viscosity μ at depths greater than 150 km is 10^{17} m^2 s^{-1} and is reduced by a factor of 50 above this interface. The Rayleigh number is calculated by reference to the high viscosity. Heating is from below with a temperature contrast ΔT of 800 °C between the bottom and the top boundaries with no slip boundary conditions. The calculation has been conducted in a box with a grid meshing of ($96 \times 64 \times 96$) and horizontal dimensions of $(3/4)^{\frac{1}{2}}d$ and $1.5d$. This aspect ratio favours the development of hexagons with a wavenumber of $\frac{4}{3}\pi$. Details of the modelling can be found in Rabinowicz *et al.* (1990).

A 3D representation of the thermal field is given in figure 1*a*. A cross section through the upper third of the box is shown on figure 1*b*, *c*. Scaling of the potential temperature has been done by assuming a value of 1280 °C in the core of the cell, according to McKenzie & Bickle (1988). This choice leads to a temperature of 1070 °C at the top of the box, compatible with the base of the lithosphere, and of 1870 °C at the upper–lower mantle transition zone, which falls in the range of estimations for the spinel–perovskite phase transition (Gasparik 1990). The heat flow transferred

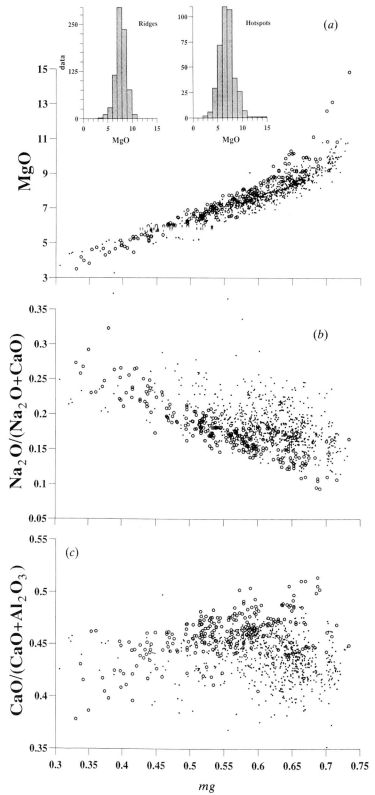

to the surface by the circulation is 60 mW m^{-2}. The plume is characterized by a relatively broad zone of gentle horizontal temperature gradient, surrounding a narrow central pipe with very steep gradients where the temperature rises by about 200 °C on a horizontal distance of a few tens of kilometres. The most striking characteristic of this thermal field is that a high temperature contrast (250–300 °C) between the core of the plume and the surrounding mantle is maintained up to the base of the upper thermal boundary layer. The flow velocity in the high viscosity layer is of the order of 70 mm a^{-1} in the plume and of 10 mm a^{-1} in the downwelling currents (not shown). This asymmetry reflects the hexagonal planform of the circulation. When it crosses the viscosity discontinuity, the ascending flow is focused towards the plume axis and the velocity increases up to 400 mm a^{-1} (figure 1*b*). Inside the LVZ, most of the plume material (about 75%) is channelled in a pipe 100 km in diameter, corresponding roughly to the thermal conduit, and will experience a large amount of adiabatic decompression due to the thinness of the upper convective boundary layer.

3. Partial melting within the plume: comparison with the composition of hotspot lavas

The solidus envelope for dry peridotite, determined experimentally by Takahashi (1986, 1990) and Herzberg *et al.* (1990), is shown on figure 1*b*, assuming an overlying lithosphere thickness of 20 km, which represents a situation of an on ridge hotspot like Iceland. Melting initiates at a depth of about 150 km, i.e. close to the base of the LVZ. Assuming no extraction, the melt fraction can easily be computed from this thermal structure. However, the melt fraction as a function of P, T is still not well determined at high pressure. We prefer to consider the composition of the residue of melting for which there are more reliable experimental constraints. The cpx-out and opx-out boundaries according to Takahashi (1986) are reported in figure 1*b, c*. In an area about 100 km in width at the top of the plume, the residue of melting should be dunitic, which corresponds to the production of highly magnesian melts ([MgO] > 25%). About 75% of the plume material is channelled into this shallow high temperature region, and most of the melt erupted at hotspots is likely to be extracted from the zone, at the base of the lithosphere, where high flow rates are realized and where the vertical flow rotates abruptly to the horizontal (Watson & McKenzie 1991). Massive eruptions of picritic melts should therefore be observed at hotspot volcanoes. It can of course be argued that the melt will separate from the plume as soon as the melt fraction exceeds several percent. Accordingly, the melts erupted at hotspots should result from lower degrees melting, but at high pressure. As the solidus composition at pressures around 5 GPa is komatiitic (Herzberg *et al.* 1990), this scenario also predicts the eruption of highly magnesian lavas.

To compare the predictions of the plume model with melt compositions observed at hotspot volcanoes, we have compiled about 1200 published analyses of glasses or, when not available, of aphyric lavas (phenocrysts content less than 1%). These analyses come from the worldwide oceanic ridges system and from several hotspots, mostly Hawaii and Iceland, where large volumes of tholeiitic basalts are erupted.

Figure 2. Variation diagrams for mid-ocean ridge basalts (dots) and hotspot (mainly Hawaii and Iceland) tholeiites (open circles). Only glasses and aphyric lavas are included in the data base (references available on request).

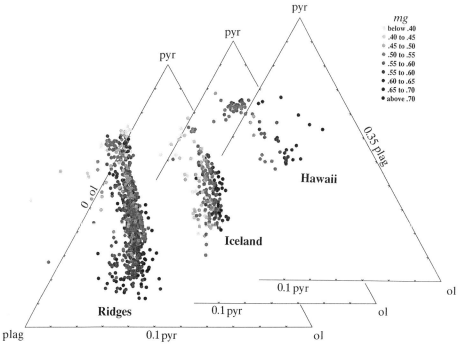

Figure 3. Plot of the same data in triangular diagrams following the projection of Bryan & Dick (1982); ol, normative olivine; pyr, normative clino- + ortho- pyroxene; plag, normative plagioclase.

The diagrams presented in figures 2 and 3 are just devoted to illustrate some basic similarities and differences between MORBs and hotspot tholeiites that any model should reproduce. Only trends of general application are presented here for the sake of discussion. An exhaustive analysis of major element chemistry of basalts can be found in Langmuir *et al.* (1992) and in Albarède (1992).

Figure 2*a* shows that the range of MgO contents of MORBs and of hotspot lavas (melts) overlap. In both environments, the average MgO content is of the order of 7%. Maximum MgO contents for ridges and hotspots are 11% and 15% respectively, but it has to be stressed that MgO contents of more than 11% are very uncommon (Clague *et al.* 1991). On figure 3, it appears that the shallow differentiation trends for ridges, Iceland and Hawaii are increasingly dominated by olivine crystallization, a characteristic which supports the view that the Hawaiian parental melt is significantly more mafic than that of MORBs, and that the one of Iceland is intermediate. However, high degree of melting followed by huge amount of low pressure fractionation cannot be invoked to explain the MgO content of hotspot tholeiites similar to those of MORBs, because the range of *mg* is similar in both environments (figure 2*a*). The slight Fe enrichment of hotspot tholeiites compared to MORBs (lower *mg* for the same MgO content, figure 2*a*) can be attributed to slightly higher amount of fractionation, but also to higher pressure of melting or to source heterogeneity (Klein & Langmuir 1987; Albarède 1992). High amount of olivine fractionation is also in contradiction with minor and trace element data (Budhan & Schmitt 1985; Feigenson 1986). Picritic lavas are observed at hotspot volcanoes, although in small amount compared to tholeiites (Nicholls & Stout 1988). However, these lavas are, as a rule, highly olivine phyric and their high MgO content reflects likely some olivine accumulation process. Such picrites are also sampled along mid-oceanic ridges far from any hotspot. Their occurrence cannot be used as an evidence for very high degrees of melting at hotspots.

For a given *mg*, the Na/Ca ratio is sensitive to the degree of shallow melting (Allan *et al.* 1987), although this character could be attributed to source heterogeneity (Albarède 1992). Figure 2*b* shows that hotspot and ridge data have overlapping variation fields for this character also and define similar trends, in spite of a higher scatter toward high Na/Ca values (low degrees of melting) is observed for ridges.

The most primitive (high *mg*) hotspot tholeiites have significantly higher Ca/Al ratio than MORBs (figure 2*c*). This difference results in a shallow differentiation trend which appears more dominated by olivine and clinopyroxene fractionation (positive slope) at hotspots than at ridges, where the trend is rather controlled by olivine and plagioclase fractionation (negative slope). This is one example among others of geochemical trends which differ in hotspot and ridge environments.

According to this brief overview, melts produced at hotspots have a composition which has been buffered in a field similar to that of MORBs when they reach the surface. No evidence of very high degree of melting, as predicted by plume models, can be found in melts erupted at hotspot volcanoes. However, in spite of these common character, significant differences appear in the differentiation trends of hotspot and ridge tholeiites and in trace elements patterns (not shown here) which indicate that different processes are involved in the genesis of the two suites. As shown by Langmuir *et al.* (1992) and Albarède (1992), these contrasting trends cannot be explained by shallow (intracrustal) differentiation alone but require differences in the high-pressure processes. In the next section, we propose a possible scenario which might account for some of these data.

4. Discussion: consequence of melt migration within the melting region of the plume

The lack of high MgO liquids at hotspot volcanoes can be accounted for in various ways. For example, Watson & McKenzie (1991) show, in the frame of a constant mantle viscosity convective model, that the melt composition averaged on the entire melting region of the plume can fit the parental melt composition assumed for Hawaiian tholeiites by adjusting the thickness of the overlying lithosphere. This model is attractive as the lithospheric thickness they found (about 70 km) is compatible with geophysical observations at Hawaii. However, it does not account for the lack of systematic variation in melt composition with lithospheric thickness from one hotspot to the next which suggests that processes independent of the amount of adiabatic decompression control the melt composition. A thick lithospheric lid obviously cannot be invoked at on-ridge hotspots.

Another plausible explanation has been proposed by Stolper & Walker (1980) who suggest that the lack of picritic liquids at the surface is just a consequence of their high density. In that frame, low pressure crystallization of olivine, clinopyroxene and plagioclase at a pseudo-eutectic point might account for the homogeneity of eruption temperatures. However, it seems that such a 'petrological buffering' cannot be invoked in the case of the most primitive melts erupted at ridges and hotspots, as they were crystallizing one or two phases only at the time they were frozen. It means that the low temperature region of the phase diagram was never reached by these melts. For instance, clinopyroxene is observed to join olivine and plagioclase as a phenocryst in the case of highly evolved MORBs only ($0.45 < mg < 0.60$) (Hess 1989).

The wide scatter of data in variation diagrams of figures 2 and 3 is a further evidence that an invariant point was reached neither by MORBs nor by OIBs.

Here, we propose an explanation to the low eruption temperature of hotspot lavas based on the possible behaviour of the first melt fractions produced in a mantle plume. Observations in mantle outcrops show that melt migration does not involve homogeneous intergranular porous flow on distances exceeding a few metres; compaction of a crystal mush (the likely physical state of a partially molten peridotite when it crosses its solidus) induces strong lateral porosity variations and causes the redistribution of the melt in a vein network whose preferred orientation is conditioned by the local stress field (Ceuleneer & Rabinowicz 1992). Classical vein thickness (E), spacing (f) are respectively 10–100 mm and 1–100 m. The size of the veined region is typically 1–10 km. This observation is corroborated by the theoretical model of Stevenson (1989) which show that in the partly molten mantle, where the compaction length (l) is estimated to several hundred metres, the melt is distributed in regions of high melt/rock ratio (veins) oriented parallel to the maximum compressive stress (σ_1) whose spacing (f) is proportional to $(lr)^{-\frac{1}{2}}$ where r is the grain radius of mantle rocks, i.e. f is of the order of 1 m. Although they are preferentially oriented along the σ_1 direction, the close spacing of the veins compared to their length implies connectivity of the veins. Furthermore, connection of the veins will eventually lead to their coalescence and will trigger the development of extremely long veins (Sleep 1988). It results that a network of veins oriented along the σ_1 direction and extending across the whole partially molten region will develop. The effective permeability K of this system along the σ_1 direction is given by the Hele–Shaw relation:

$$K = E^3/12 f. \tag{2}$$

In the case of 5% melt fraction distributed in veins with $f = 1$ m, the permeability K is about 10^{-5} m^2.

From the base of the melting region to depths of around 80–100 km, σ_1 is sub-horizontal; its dip is of the order of 10–20° toward the axis of the plume (figure 1b). The extension direction (not shown) is sub-vertical. Accordingly, a network of sub-horizontal veins radiating away from the plume axis will develop in this lower part of the melting region. Assuming this system is interconnected, buoyancy forces could lead to initiation of hydrofracturing which would result in rapid melt migration to the surface (Nicolas 1986). However, due to the small density contrast between melt and solid at such pressures (Herzberg 1987), and to the sub-horizontal attitude of the veins, hydrofracturing can only occur if interconnection extends for a few hundred kilometres at least which exceeds the dimension of the melting region! The melt will thus remain trapped within the vein network. The conclusion that hydrofracture is unlikely even in the shallow asthenosphere is corroborated by observations in ophiolites which show that dyking is absent from zones of palaeovertical mantle flow and is typically an off-axis mechanism. Melt percolation in a network of close spaced veins is found to be the dominant migration mechanism in mantle diapirs (Ceuleneer & Rabinowicz 1992).

If hydrofracture does not occur, the strong temperature contrast between the centre and the periphery of the plume will probably induce convection of the melt within the vein network. Due to the geometry of the system, flow can be described as Hele–Shaw convection in an inclined layer. The porous Ra number is:

$$Ra_p = \alpha_1 \Delta T g \sin (\Theta) LK/\kappa\nu_1, \tag{3}$$

where α_1 and ν_1 are the coefficient of thermal expansion and the viscosity of the liquid, Θ the dip of σ_1, L is the interconnection length along the σ_1 direction and ΔT is the temperature contrast between the centre and the periphery of the melting region. For high pressure melts, α_1 is of the order of $10^{-4}\,°C^{-1}$ and ν_1 of the order of $10^{-3}\,m^2\,s^{-1}$ (Kushiro 1980; Herzberg 1987).

Figure 1*b*, shows that L may reach a value of about 50 km at depths around 100 km, if interconnection is achieved from the centre to the periphery of the melting region; Θ is of the order of $20°$ and the thermal contrast is of the order of 150 °C. This lead to an estimation for Ra_p of the order of 2.5×10^7. This is a huge value for porous flow convection, and shows that a large amount of heat can be advected by melt circulation. The effective diffusivity of heat within the melting region can be estimated using

$$\kappa_{ef} = \kappa Nu, \tag{4}$$

where Nu is the Nusselt number of the porous flow convection. Recent estimations of the relation $Nu f(Ra)$ for large values of Ra give (Lister 1990):

$$Nu = 0.45\,Ra^{0.55}. \tag{5}$$

Accordingly, values for Nu greater than 10^3 are possible, and will promote the rapid homogenization of the plume temperature. A characteristic time of attenuation of the thermal contrast existing in a finite circular tube is given by (see Carslaw & Jaeger 1978, p. 195):

$$a^2/\kappa_{ef}\beta^2, \tag{6}$$

where $\beta = 2.4048$, is the smallest root of the Bessel function of order 0. According to the above estimates, temperature homogenization could be achieved in less than 2×10^4 years, the time needed by the mantle plume to rise by about 10 km. This reasoning shows that complete homogenization of the plume temperature can be achieved due to porous flow convection in the vein network as soon as the melting region has reached a sufficiently large horizontal extension. We have calculated that if the temperature is fully homogenized in the thermal conduit, conservation of the heat advected by the plume requires a plume temperature exceeding the temperature of the surrounding mantle by about 50 °C.

According to observations in mantle outcrops, interconnection is achieved as soon as the melt fraction exceeds several percent which will occur at high pressure when the temperature exceeds the solidus temperature by a small amount (Takahashi 1986; Herzberg *et al.* 1990). Higher degree of melting will not be realized as the heat released by further decompression will be removed from the high temperature region of the plume by melt convection. Melt will also undergo fractional crystallization as it moves toward the periphery of the plume, which will progressively increase its fluid and LIL contents. Starting with a komatiitic liquid, the solidus composition at 5 GPa (Herzberg 1990), the melt will evolve mainly in response to crystallization of olivine and garnet: a Ca-poor garnet has been shown to follow olivine in the crystallization sequence of a komatiite around 5 GPa, while pyroxene joins these two phases later on (Herzberg *et al.* 1990). Garnet crystallization will induce some characteristic trends in both major and trace element compositions of the residual liquid, like a high Ca/Al and a high La/Yb. It is tentatively proposed that, in the lower part of the melting region (5–3 GPa), the melt circulating in the drainage network acquires progressively a kimberlitic composition, which is intermediate between an ultrabasic and an alkalic end member and displays some of the geochemical characters

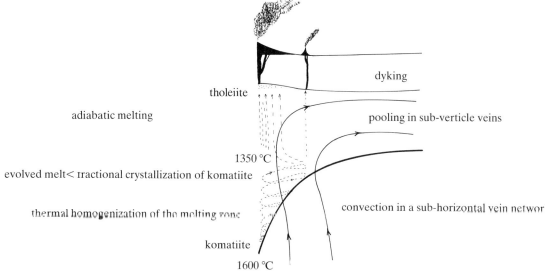

Figure 4. Sketch illustrating solid-state flow trajectories (solid lines) and melt flow trajectories (dashed lines) in the melting region of the convective plume. The heavy curve is the solidus.

mentioned above (Wyllie 1988). The progressive increase of the fluid content can promote remelting of the wall rock by lowering its solidus temperature (Wyllie 1988), and will prevent complete 'freezing' of the melt drainage system. Here again, a kimberlitic melt will be produced as shown by high pressure melting experiments in conditions of CO_2 saturation (Canil & Scarfe 1990). In the case of a very thick lithospheric lid (craton) overlying the plume, this melt may reach the surface. In the case of a thinner lithosphere, the composition of the melt circulating in the vein network will continue to evolve in response to both fractional crystallization and introduction of lower pressure melt fractions at the periphery of the plume (figure 4). Among other effects, these two processes will contribute to lower the Mg content of the melt and may move its composition toward the one of an alkalic lava. Finally, where there is no obstacle to further decompression (on-ridge hotspot or deeply eroded lithosphere), a large volume of the plume will melt as the solidus envelope enlarges (figure 4). However, the temperature in the melting region is buffered to a value close to the average mantle potential temperature, and the plume will produce a tholeiitic melt. Since the maximum compressive stress is now sub-vertical, melt fractions will collect in steeply dipping veins and eventually cracks which will trigger rapid melt migration toward the surface. The lavas erupted at hotspots are likely to result from the mixing in various proportion of these tholeiites with the residual melts produced in the lower part of the melting zone. This process may account for some petrological peculiarities (relative to MORBs of similar *mg*) of hotspot tholeiites, like their 'enriched' signature and other geochemical characters discussed above.

There are two assumptions which are critical for the present model: (i) partial melting of the plume is associated with a viscosity drop which induces the focusing of the mantle flow and, consequently, a sub-horizontal orientation of the maximum compressive stress; (ii) that intergranular partial melt is not stable in intergranular position but migrates to form a vein network. These two points are among the most firmly established conclusions derived from structural observations in mantle outcrops. There is of course no direct evidence that the interplay between solid-state flow and melt migration follows the same modalities at 150 km depth. However, if both assumptions are valid, they imply that melt differentiation and thermal homogenization will take place in the melting region of mantle plumes.

This work has greatly benefited from discussions with Peter Kelemen and Claude Dupuy, and from a constructive review of Dan McKenzie. We are very indebted to Philippe Vidal for his repeated encouragements to pursue this direction of research. This is a contribution CNRS–INSU, programme DBT, thème 'Instabilités', no. 476.

References

Albarède, F. 1992 How deep do common basaltic magmas form and differentiate? *J. geophys. Res.* **27**, 10997–11009.

Allan, J. F., Batiza, R. & Lonsdale, P. 1987 Petrology of lavas from seamount flanking the East Pacific Rise axis at 21° N: implications concerning the mantle source composition for both seamount and adjacent EPR lavas. In *Seamounts, islands and atolls*, pp. 255–282. AGU Monographs.

Bryan, W. B. & Dick, H. J. B. 1982 Contrasted abyssal basalt liquidus trends: evidence for mantle major element heterogeneity. *Earth planet. Sci. Lett.* **58**, 15–26.

Budhan, J. R. & Schmitt, R. A. 1985 Petrogenetic modeling of Hawaiian tholeiitic basalts: a geochemical approach. *Geochem. cosmochim. Acta* **49**, 67–87.

Canil, D. & Scarfe, C. 1990 Phase relations in peridotites + CO_2 systems to 12 Pa: implications for the origine of kimberlite and carbonate stability in the Earth's upper mantle. *J. geophys. Res.* **95**, 15805–15816.

Ceuleneer, G. & Rabinowicz, M. 1992 *Mantle flow and melt migration beneath oceanic ridges: models derived from observations in ophiolites* (ed. D. Blackman & J. Phipps Morgan). AGU Monographs. (In the press.)

Ceuleneer, G., Rabinowicz, M., Monnereau, M., Cazenave, A. & Rosemberg, C. 1988 Viscosity and thickness of the sublithospheric low-viscosity zone: constraints from geoid and depth over oceanic swells. *Earth planet. Sci. Lett.* **89**, 84–102.

Clague, D. A., Weber, W. S. & Dixon, J. E. 1991 Picritic glasses from Hawaii. *Nature, Lond.* **353**, 553–556.

Feigenson, M. D. 1986 Constraints on the origin of Hawaiian lavas. *J. geophys. Res.* **91**, 9383–9393.

Gasparik, T. 1990 Phase relations in the transition zone. *J. geophys. Res.* **95**, 15751–15769.

Helz, R. T. 1987 Diverse olivine types in lava of the 1950 eruption of Kilauea volcano and their bearing on eruption dynamics. *U.S. Geol. Survey. Prof. Papers* **1350**, 691–722.

Herzberg, C. T. 1987 Magma density at high pressure. Part 1. The effect of composition on the elastic properties of silicate liquids. *Geochem. Soc. Spec. Publ.* **1**, 25–46.

Herzberg, C., Gasparik, T. & Sawamoto, H. 1990 Origin of mantle peridotite: constraints from melting experiments to 16.5 GPa. *J. geophys. Res.* **95**, 15779–15804.

Hess, P. C. 1989 *Origins of igneous rocks*. (336 pages.) Harvard University Press.

Jakobsson, S. P., Jonsson, J. & Shido, F. 1978 Petrology of the western Reykjanes peninsula, Iceland. *J. Petrol.* **19**, 669–705.

Klein, E. M. & Langmuir, C. H. 1987 Global correlations of ocean ridge basalt chemistry with axial depth and crustal thickness. *J. geophys. Res.* **92**, 8089–8115.

Kushiro, I. 1980 Viscosity, density, and structure of silicate melts at high pressure, and their petrological applications. In *Physics of magmatic processes* (ed. R. B. Hargraves), pp. 93–120. Princeton University Press.

Langmuir, C., Klein, E. & Plank, T. 1992 *Petrological constraints on melt formation and segregation beneath MOR* (ed. D. Blackman & J. Phipps-Morgan). AGU Monograph. (In the press.)

Lister, C. R. B. 1990 An explanation for the multivalued heat transport found experimentally for convection in a porous medium. *J. Fluid Mech.* **214**, 287–320.

McKenzie, D. P. & Bickle, M. J. 1988 The volume and composition of melt generated by extension of the lithosphere. *J. Petrol.* **29**, 625–679.

Nicolas, A. 1986 Melt extraction model based on structural studies in mantle peridotites. *J. Petrol.* **27**, 999–1022.

Nicholls, J. & Stout, M. Z. 1988 Picritic melts in Kilauea – evidence from the 1967–1968 Halemaumau and Hiiaka eruptions. *J. Petrol.* **29**, 1031–1057.

Rabinowicz, M., Ceuleneer, G., Monnereau, M. & Rosemberg, C. 1990 Three-dimensional models of mantle flow across a low viscosity zone: implications for hotspot dynamics. *Earth planet. Sci. Lett.* **99**, 170–184.

Robinson, E. M., Parsons, B. & Daly, S. F. 1987 The effect of a shallow low viscosity zone on the apparent compensation of mid-plate swells. *Earth planet. Sci. Lett.* **82**, 335–348.

Sleep, N. H. 1988 Tapping of melt by vein and dykes. *J. geophys. Res.* **93**, 10255–10272.

Stevenson, D. J. 1989 Spontaneous small-scale melt segregation in partial melts undergoing deformation. *Geophys. Res. Lett.* **16**, 1067–1070.

Stolper, E. M. & Walker, D. 1980 Melt density and the average composition of basalt. *Contrib. Mineral. Petrol.* **74**, 7–12.

Takahashi, E. 1986 Melting of a dry peridotite KLB-1 up to 14 GPa: implications on the origin of peridotitic upper mantle. *J. geophys. Res.* **91**, 9367–9382.

Watson, S. & McKenzie, D. P. 1991 Melt generation by plumes: a study of Hawaiian volcanism. *J. Petrol.* **32**, 501–537.

White, R. S. & McKenzie, D. P. 1989 Magmatism at rift zones: the generation of volcanic continental margins and flood basalts. *J. geophys. Res.* **94**, 7685–7729.

Wyllie, P. J. 1988 Magma genesis, plate tectonics and chemical differentiation of the Earth. *Rev. Geophys.* **26**, 370–404.

Estimates of mantle thorium/uranium ratios from Th, U and Pb isotope abundances in basaltic melts

By R. Keith O'Nions and Dan McKenzie

Institute of Theoretical Geophysics, Department of Earth Sciences, Downing Street, Cambridge CB2 EQU, U.K.

The relationship between the abundances of Th, U and Pb isotopes in basalt melts and the [Th/U] ratio of their source is assessed. A simple melting model is used to show that whereas the activity ratio (^{230}Th/^{232}Th) in the initial melt before extraction is equal to the bulk source ratio, that in the extracted melt may be higher. The difference depends upon the rate of melting relative to the half-life of ^{230}Th (73 ka). Only when the rate is fast compared to this half life will (^{230}Th/^{232}Th) in the extracted melt provide a correct estimate of [^{232}Th/^{238}U] in the source and therefore of its [Th/U] ratio. This is normally not the case for MORB, and a better estimate of source [Th/U] ratio is derived from [^{232}Th/^{238}U] ratio in the basalt, which does not depend upon the rate of melting. Available data for MORB glasses give a best estimate for their source [Th/U] = 2.58±0.06. This value is less than both that of the bulk Earth of 3.9±0.1, and of the source of plume basalts from Iceland and Hawaii, which are 3.3 and 3.2 respectively.

These estimates contrast with the [^{232}Th/^{238}U] ratio required to produce the radiogenic {^{208}Pb/^{206}Pb} atomic ratio of MORB over 4.55 Ga. This averages 3.8 and is little different from the average derived from Pb-isotopes in plume basalts. These observations are most easily reconciled if Th, U and Pb are efficiently stripped from the mantle by melting and have a residence time there of ⩽ 1 Ga. The [Th/U] ratio of 2.6 for the upper mantle requires melt fractions of ⩽ 1% to be involved in transferring U and Th from this region into the continents. Such melt fractions are present in subduction zones and in the source regions of continental alkali basalts.

1. Introduction

In recent years the physical controls on the generation and extraction of mantle partial melt have become much better understood (see Spiegelman, this symposium). This understanding has resulted in renewed interest in the information that melt compositions provide about the depth and amount of melting in the Earth. Th, U and Pb are all highly incompatible elements, which partition strongly into the melt and are therefore expected to be depleted in the mantle residues even when the melt fractions are small. ^{232}Th, ^{235}U and ^{238}U decay to Pb via a series of intermediate daughter products of which some, notably ^{230}Th ($t_{\frac{1}{2}} = 73$ ka) and ^{226}Ra ($t_{\frac{1}{2}} = 1.6$ ka) are particularly useful in studies of melting and melt separation (Allègre & Condomines 1982; Cortini 1984; McKenzie 1985; Williams & Gill 1989). Also U and Th have decayed to stable isotopes of Pb over some 4.5 Ga and Pb isotope ratios in the Earth therefore record its [Th/U] ratio. It has been commonplace (Galer & O'Nions 1985; Condomines *et al.* 1988; Goldstein *et al.* 1991; Sigmarsson *et al.* 1992)

Phil. Trans. R. Soc. Lond. A (1993) **342**, 65–77

Printed in Great Britain

to assume that the $(^{230}\text{Th}/^{232}\text{Th})$ activity ratio in a melt is equal to the bulk $(^{238}\text{U}/^{232}\text{Th})$ ratio in the source region, and then to calculate [Th/U] directly from $(^{230}\text{Th}/^{232}\text{Th})$. The use of $(^{230}\text{Th}/^{232}\text{Th})$ for this purpose avoids possible Th–U fractionation during crystallization of the melt. Unfortunately the relationship between $(^{230}\text{Th}/^{232}\text{Th})$ in the melt and source [Th/U] depends critically upon the details of melting and melt-extraction, and $(^{230}\text{Th}/^{232}\text{Th})$ in the melt need not equal $(^{238}\text{U}/^{232}\text{Th})$ in the source. Therefore this approach may lead to significant errors in the estimate of mantle [Th/U].

Galer & O'Nions (1985) used different estimates of the mantle [Th/U] ratio derived from Th-, U- and Pb-isotope abundances in mid-ocean ridge basalts to deduce important information about the circulation pattern in the mantle. In particular they argued that the residence time of Th, U and Pb in the upper mantle must be considerably less than the age of the Earth. This geochemical argument is one of the principal reasons for believing that there is only limited exchange of material between the upper and lower mantle. However, their argument depended on the use of $(^{230}\text{Th}/^{232}\text{Th})$ to estimate the present value of [Th/U] in the upper mantle, using the assumption that the $(^{230}\text{Th}/^{232}\text{Th})$ activity ratio in a melt is equal to the bulk $(^{238}\text{U}/^{232}\text{Th})$ ratio in the source region. Since this assumption is not correct, and since the constraint that Galer & O'Nions (1985) imposed on mantle circulation is so central to any attempt to understand mantle convection, it is essential to re-examine their whole argument. This we do below, and show that Galer & O'Nions's (1985) results are correct, even though one part of their argument is not.

2. Th, U and Pb isotopes in melts

The isotopes of interest here are the two long-lived parents, ^{238}U ($t_{\frac{1}{2}} = 4.47$ Ga) and ^{232}Th ($t_{\frac{1}{2}} = 14$ Ga), and their respective stable daughter products, ^{206}Pb and ^{208}Pb, together with ^{230}Th ($t_{\frac{1}{2}} = 73$ ka), the radioactive intermediate daughter in the ^{238}U decay series. At radioactive equilibrium in the mantle $(^{238}\text{U}) = (^{230}\text{Th})$ and therefore $(^{230}\text{Th}/^{232}\text{Th}) = (^{238}\text{U}/^{232}\text{Th})$, where ( ) are used to denote the activities of the isotopes concerned, [ ] the concentration by weight, and { } atomic ratios. The presence or absence of this equilibrium in mantle partial melts has been extensively documented and discussed (see, for example, Condomines *et al.* 1988). In principle there are three different measures of the [Th/U] ratio in the mantle, which may be derived from $(^{232}\text{Th}/^{230}\text{Th})$, $[^{232}\text{Th}/^{238}\text{U}]$, or $\{^{208}\text{Pb}/^{206}\text{Pb}\}$ ratios.

The effects of partial melting on these radioactive and stable isotopes are now considered using the melting model described by McKenzie (1985), which is similar to dynamic melting model of Langmuir *et al.* (1977) and of Williams & Gill (1989). Melting is assumed to occur at a constant rate as pressure decreases and the melt produced is in chemical and radioactive equilibrium with the matrix. The melt fraction by volume ϕ remains constant throughout the melting region, from which melt is extracted continuously and instantaneously. Melt once extracted is unable to re-equilibrate with the matrix and is mixed completely with melts extracted at other depths. The equations governing the conservation of ^{238}U, ^{232}Th and ^{230}Th are (McKenzie 1985)

$$\text{d}a_{\text{p}}/\text{d}t = -\alpha_{\text{p}}\,a_{\text{p}}, \tag{1}$$

$$\text{d}a_{\text{r}}/\text{d}t = -\alpha_{\text{d}}\,a_{\text{r}}, \tag{2}$$

and
$$\text{d}a_{\text{d}}/\text{d}t = -\alpha_{\text{d}}\,a_{\text{d}} + \lambda_{\text{d}}((F_{\text{d}}/F_{\text{p}})\,a_{\text{p}} - a_{\text{d}}), \tag{3}$$

where $\lambda_d = \lambda_{230}$, and a_p and a_d are the true activities of the parent ^{238}U and daughter ^{230}Th, but $a_r = \lambda_{230}[^{232}$Th], is not. It is convenient to define a_r in this way so that

$$a_d/a_r = [^{230}\text{Th}/^{232}\text{Th}] \equiv R. \tag{4}$$

F is given by
$$F = (K\rho_s(1-\phi)/\rho_f\phi+1)^{-1}, \tag{5}$$

where K is a partition coefficient, ρ_s and ρ_f are the densities of the solid and the melt, ϕ is the melt fraction present in the source by volume, and

$$\alpha = F\Gamma/\rho_f\phi, \tag{6}$$

where Γ is the melting rate. Equations (1)–(3) assume that $1 \gg K_p$, K_d, and therefore all expressions obtained below are accurate only to $O(K_d)$ or $O(K_p)$, whichever is larger. Throughout the discussion below we neglect the difference in density between the solid and the melt and therefore take $\rho_s = \rho_f = \rho$. We also assume that $1 \gg \phi$, when

$$\alpha = \Gamma/\rho(K+\phi) \tag{7}$$

and
$$F = \phi/(K+\phi). \tag{8}$$

The solutions to equations (1)–(3) are (McKenzie 1985)

$$a_p(t) = a_p(0)\,e^{-\alpha_p t}, \tag{9}$$

$$a_r(t) = a_r(0)\,e^{-\alpha_d t}, \tag{10}$$

$$a_d(t) = a_d(0)\,e^{-(\lambda_d+\alpha_d)t} + a_p(0)\,[\lambda_d F_d/(\lambda_d+\alpha_d-\alpha_p)]\,(e^{-\alpha_p t} - e^{-(\lambda_d+\alpha_d)t}). \tag{11}$$

It is generally assumed that the $[^{230}\text{Th}/^{232}\text{Th}]$ ratio, R, in the melt in equilibrium with the matrix is equal to that in the bulk source. This result is easily proved. The mean concentration $\bar{c}$ of ^{238}U, ^{230}Th and ^{232}Th of the combined matrix and melt are

$$\bar{c}_p = c_f^p\,\phi\rho/F_p, \quad \bar{c}_d = c_f^d\,\phi\rho/F_d, \quad \bar{c}_r = c_f^r\,\phi\rho/F_d, \tag{12}$$

respectively, where c_f is the concentration in the melt. Hence the bulk source ratio $[^{230}\text{Th}/^{232}\text{Th}]$, R_B, is given by

$$R_B \equiv \frac{\bar{c}_d}{\bar{c}_r} = \frac{\phi\rho c_f^d/F_d}{\phi\rho c_f^r/F_d} = \frac{c_f^d}{c_f^r}. \tag{13}$$

Therefore at the onset of melting

$$R_B(0) = \bar{c}_d(0)/\bar{c}_r(0) = a_d(0)/a_r(0) = R(0). \tag{14}$$

Furthermore, since $(^{230}$Th$)$ and $(^{238}$U$)$ are equal in radioactive equilibrium, it follows that

$$\lambda_d\bar{c}_d = \lambda_p\bar{c}_p. \tag{15}$$

Dividing by $\lambda_d\bar{c}_r$ gives
$$R_B(0) = (\lambda_p/\lambda_d)\,\bar{c}_p/\bar{c}_r. \tag{16}$$

If therefore $R_B(0)$ or $R(0)$ can be determined, so can $\bar{c}_p/\bar{c}_r$, the [U/Th] ratio of the source. Substitution for $\bar{c}$ in equation (15) using (12) gives

$$(c_f^d(0)/F_d)\,\lambda_d = (c_f^p(0)/F_p)\,\lambda_p. \tag{17}$$

Therefore
$$a_p(0)/a_d(0) = F_p/F_d \tag{18}$$

and hence from equation (14)

$$G_f(0) = (F_p/F_d)\,R(0) = (F_p/F_d)\,R_B(0), \tag{19}$$

where
$$G_f = a_p/a_r = (\lambda_p/\lambda_d)\,[^{238}U/^{232}Th]_f. \tag{20}$$

If melt is now extracted from the whole column that is melting, and accumulated for a time t, the $[^{230}Th/^{232}Th]$ ratio in the accumulated melt is given by

$$R(t) = \int_0^t a_d(t')\,dt' \Big/ \int_0^t a_r(t')\,dt'. \tag{21}$$

In the limit as $t \to \infty$ integration of equations (10) and (11) gives

$$R(\infty) = \frac{a_d(0)}{\lambda_d + \alpha_d}\left(1 + \frac{\lambda_d\,F_d}{\alpha_p\,F_p}\left(\frac{a_p(0)}{a_d(0)}\right)\right)\Big/\left(\frac{a_r(0)}{\alpha_d}\right). \tag{22}$$

Substitution of equations (14) and (18) into (22) gives

$$R(\infty) \approx R_B(0)\,(\alpha_d/(\lambda_d + \alpha_d))\,(1 + \lambda_d/\alpha_p). \tag{23}$$

Equation (23) shows that $R(\infty)/R_B(0)$, the ratio of $[^{230}Th/^{232}Th]$ is the accumulated melt to that in the source, depends upon λ_d, α_d and α_p, and therefore on both the melting rate Γ and the ^{230}Th decay rate, as well as on the partition coefficients K_p and K_d.

Consider first *slow* melting where $\lambda_d (= 9.5 \times 10^{-6}\,a^{-1}) \gg \Gamma/\rho\phi$, and therefore $\lambda_d \gg \alpha_d,\,\alpha_p$. Substitution of equation (7) into (23) then gives

$$R(\infty)/R_B(0) \approx \alpha_d/\alpha_p = (K_p + \phi)/(K_d + \phi). \tag{24}$$

Thus whenever $K_p > K_d$, $R(\infty)/R_B(0) > 1$, and $[^{230}Th/^{232}Th]$ in the extracted melt will exceed that in both the initial melt and the bulk source. $[^{230}Th/^{232}Th]$ in the extracted melt is often used as a direct measure of the $[Th/U]$ source ratio, but in fact it will underestimate it by an amount which depends on K_p, K_d and the melt fraction ϕ present during melting. For this reason the estimates of $[Th/U]$ in the MORB source made by Galer & O'Nions (1985) that were based on $[^{230}Th/^{232}Th]$ are likely to be in error.

If on the other hand *rapid* melting takes place, and α_p, $\alpha_d \gg \lambda_d$, then equation (23) simplifies to $R(\infty) = R_B(0)$ and the $[^{230}Th/^{232}Th]$ in the extracted melt equals that in the bulk source. These results were first obtained by Williams & Gill (1989).

Equation (14) shows that $[^{230}Th/^{232}Th]$ in the initial melt is equal to that in the bulk source. The same result is not in general true for $[^{238}U/^{232}Th]$. Equation (19) shows that the $[^{238}U/^{232}Th]$ in the initial melt is not equal to the bulk ratio, R_B, unless $F_p = F_d$, that is $K_p = K_d$, rather than $K_p > K_d$, which is the expected situation.

The $[^{238}U/^{232}Th]$ ratio in melt extracted during a time t is proportional to G_f, where

$$G_f(t) = \int_0^t a_p(t')\,dt' \Big/ \int_0^t a_r(t')\,dt'. \tag{25}$$

In the limit as $t \to \infty$ integration gives

$$G_f(\infty) = \alpha_d\,a_p(0)/\alpha_p\,a_r(0) = (\alpha_d/\alpha_p)\,G_f(0). \tag{26}$$

Substitution for α_d and α_p from equation (7) gives

$$G_f(\infty) = [(K_p + \phi)]/(K_d + \phi)]\,G_f(0), \tag{27}$$

where the term in square brackets is $R(\infty)/R_B(0)$ (see equation (24)). Even though $1 \gg K_p,\,K_d,\,\phi$, this term is the ratio of these small quantities, and may therefore differ substantially from unity. If $K_p > K_d$ [^{238}U/^{232}Th] in the accumulated extracted melt will exceed the ratio in the melt at the start of extraction. This result is to be expected, because U will be retained by the matrix more than Th during the initial melting, but will eventually all be extracted as melting proceeds.

Substitution in equation (27) for $G_f(0)$ from equation (19) and for F_d and F_p from (8) gives

$$G_f(\infty) = R_B(0). \tag{28}$$

Substitution for $R_B(0)$ and $G_f(\infty)$ using (16) and (20) then gives

$$[^{238}\text{U}/^{232}\text{Th}]_f = \bar{c}_p/\bar{c}_r. \tag{29}$$

[^{238}U/^{232}Th] in the accumulated extract will therefore give the bulk source value. Unless other complicating factors become important, such as alteration or low pressure crystal fractionation, [^{232}Th/^{238}U] in the melt should be a reliable measure of [Th/U] in the bulk source, and is to be preferred for this purpose to estimates based upon (^{230}Th/^{232}Th).

For completeness the isotopes of Pb are discussed briefly. ^{232}Th, ^{235}U and ^{238}U undergo radioactive decay to ^{208}Pb, ^{207}Pb and ^{206}Pb respectively. These Pb isotopes, together with ^{204}Pb which does not have a radioactive parent, are all stable. They all have a straightforward behaviour during melting because the rate of change of their concentrations due to radioactive decay is too slow for any measurable change to occur during melt extraction. In relationship to the equations derived above, melting is always rapid in the case of Pb. Therefore, if Pb is in chemical equilibrium in the source rock before melting starts, the relative abundances of ^{208}Pb, ^{207}Pb, ^{206}Pb and ^{204}Pb are identical in the accumulated melt, the initial melt and the source, regardless of what assumptions are made about K_{Pb} and about whether or not the melt itself is in chemical equilibrium with the matrix. Reconstruction of the time averaged value of [Th/U] for a source from Pb-isotope ratios is therefore straightforward. Because Pb is highly incompatible during MORB melt production, and $K_{Pb} \approx K_U \approx K_{Th} \ll 1$, Pb, like U and Th, will be almost completely removed by melts, even when the melt fractions are small.

3. Estimation of mantle [Th/U]

The [Th/U] ratio for mantle source regions may be estimated in principle from the Th-, U-, and Pb-isotope abundances in basalts extracted from them. Such estimates are now made from the Th, U and Pb data available for basalts in the light of the above considerations of their behaviour during melting. To avoid the effects of alteration on Th- and U-isotope abundances, only those analyses made on basalt glasses are used to estimate [Th/U]. Published data of this sort are strongly biased to samples obtained from the East Pacific ridges. Of the analyses from ocean ridge basalts summarized in table 1, only the set published by Jochum *et al.* (1983) contain a large number of Atlantic samples.

From the above arguments it is clear that (^{230}Th/^{232}Th) of basalts should not be used to estimate mantle [Th/U]. Rather [Th/U] ratios of the basalts themselves are likely to provide a more reliable estimate. This is evident in table 1, where four sets of ocean ridge basalt analyses, for which Th- and U-isotope data exist, each have [^{232}Th/^{238}U] ratios calculated from (^{230}Th/^{232}Th) that are less than the measured [^{232}Th/^{238}U] ratios of the basalts themselves. This result is to be expected, because (^{230}Th/^{238}U) is greater than 1.0 in each of the four data-sets. Measured [^{232}Th/^{238}U]

Table 1. Th–U *isotopes in* MORB *and ocean islands*

(All uncertainties are 1σ.)

	measured [^{232}Th/^{238}U]	measured (^{230}Th/^{232}Th)	measured (^{230}Th/^{238}U)	[^{232}Th/^{238}U] calculated from col. 2
ocean ridge basalts				
East Pacific Rise, 13° N ($n = 9$)[a] 2.56 ± 0.29	1.31 ± 0.05	1.10 ± 0.13	2.33 ± 0.08
Juan da Fuca Ridge ($n = 16$)[b] 2.76 ± 0.28	1.30 ± 0.11	1.17 ± 0.10	2.37 ± 0.24
Gorda Ridge ($n = 11$)[b] 2.91 ± 0.23	1.19 ± 0.10	1.15 ± 0.07	2.58 ± 0.22
East Pacific Rise, 12° N–27° S ($n = 11$)[c] 2.26 ± 0.65	1.62 ± 0.42	1.07 ± 0.11	2.01 ± 0.49
Atlantic and East Pacific ($n = 18$)[d] 2.42 ± 0.39	—		
average of all oceanic ridges 2.58 ± 0.06	—	—	—
ocean island basalts				
Iceland ($n = 24$)[e] 3.27 ± 0.12	1.11 ± 0.05	1.20 ± 0.07	2.80 ± 0.13
Hawaii ($n = 13$)[f] 3.19 ± 0.33	1.05 ± 0.06	1.06 ± 0.08	2.96 ± 0.17

[a] Othman & Allègre (1990). [^{232}Th/^{238}U] ratios by TIMS, (^{230}Th/^{232}Th) by alpha-counting.
[b] Goldstein *et al.* (1991). All ratios by TIMS.
[c] Rubin & Macdougall (1988). All ratios by alpha-counting.
[d] Jochum *et al.* (1983). Ratios by spark source MS. The three samples with the largest errors were omitted from the average.
[e] Sigmarsson *et al.* (1992). [^{232}Th/^{238}U] by TIMS and (^{230}Th/^{232}Th) by alpha-counting.
[f] Newman *et al.* (1984). All ratios by alpha-counting.

ratios in ocean ridge basalts show variations that are larger than analytical errors. Though samples from the Juan da Fuca and Gorda Ridges have the highest average values (table 1), they are indistinguishable from the others at the 2σ level. A best estimate for the mantle [Th/U] ratio of 2.58 ± 0.06 ($1\sigma_m$) is obtained from all 65 samples in table 1.

With some straightforward assumptions a further estimate of mantle [Th/U] may be derived from its Pb-isotope composition. Because Pb isotopes should be identical in the mantle source, the initial and the extracted melts, the large number of high-quality Pb-isotope analyses for ridge basalts may be used with confidence. The $\{^{206,207,208}$Pb/^{204}Pb$\}$ ratios for the Pb incorporated in the Earth at the time of its formation can be estimated from meteorite analyses (Tatsumoto *et al.* 1973). The amounts of radiogenic ^{206}Pb* and ^{208}Pb* subsequently produced from parent ^{238}U and ^{232}Th in the Earth are derived from differences between the measured $\{^{206,208}$Pb/^{204}Pb$\}$ ratios in basalts and these initial Pb values. It is usual to make the simplifying assumption of closed system mantle evolution from 4.55 Ga to present day, and to term the [^{232}Th/^{238}U] ratio calculated in this way the single-stage ratio (Galer & O'Nions 1985). This ratio is usually written as an atomic, rather than a weight, ratio and is denoted as $\bar\kappa_{Pb}$, given by (Galer & O'Nions 1985)

$$\bar\kappa_{Pb} = \left\{ \frac{^{208}Pb*}{^{206}Pb*} \right\} \left(\frac{\exp(\lambda_{238}\,T) - 1}{\exp(\lambda_{232}\,T) - 1} \right), \tag{30}$$

Table 2. *Summary of [Th/U] ratios estimated from Pb isotopes (from Galer & O'Nions 1985)*
$$[^{232}\text{Th}/^{238}\text{U}]\ (\bar{\kappa}_{\text{Pb}})$$

ocean ridge basalts	
Atlantic	3.78 ± 0.07
East Pacific Rise	3.73 ± 0.06
Mid-Indian	3.89 ± 0.11
average $= 3.80$	
ocean islands	
Hawaii	3.84 ± 0.04
Iceland	3.81 ± 0.03
St Helena	3.75 ± 0.02
Australes	3.75
Ascension	3.81 ± 0.02
Azores	3.84
Bouvet	3.84
Canaries	3.92
Marquesas	3.98 ± 0.04
Guadeloupe	3.99
Tristan da Cunha	4.17
Gough	4.20
Kerguelen	4.24 ± 0.12
average $= 3.93$	

where $T = 4.55$ Ga, λ_{232} and λ_{238} are the decay constants of ^{232}Th and ^{238}U, and

$$\left\{ \frac{^{208}\text{Pb}^*}{^{206}\text{Pb}^*} \right\} = \left(\left\{ \frac{^{208}\text{Pb}}{^{204}\text{Pb}} \right\}_m - \left\{ \frac{^{208}\text{Pb}}{^{204}\text{Pb}} \right\}_I \right) \Big/ \left(\left\{ \frac{^{206}\text{Pb}}{^{204}\text{Pb}} \right\}_m - \left\{ \frac{^{206}\text{Pb}}{^{204}\text{Pb}} \right\}_I \right), \tag{31}$$

where m is the measured ratio in the basalt, and I is the initial ratio of 4.55 Ga from Tatsumoto *et al.* 1973. $\bar{\kappa}_{\text{Pb}}$ values obtained in this way are summarized in table 2. The range obtained from the Pb measurements is from 3.73 to 3.89, and is very different from the average value of 2.58 ± 0.09 derived from the $[^{232}\text{Th}/^{238}\text{U}]$ ratios in table 1.

In comparison to MORB, Th and U isotopes have been measured in basalts from comparatively few ocean islands. Though useful amounts of data are available for only Iceland and Hawaii, these are sufficient to highlight the important differences that exist between islands and the normal spreading ridges. A large amount of data for Icelandic basalts has been obtained by Sigmarsson *et al.* (1992). The average $(^{230}\text{Th}/^{238}\text{U})$ ratio is 1.20 ± 0.07 (table 1), a value similar to normal ridge basalts. Again this result means that $(^{230}\text{Th}/^{232}\text{Th})$ should not be used to estimate the source [Th/U] directly. Rather $[^{232}\text{Th}/^{238}\text{U}]$ values provide a better estimate, with $[\text{Th/U}] = 3.27 \pm 0.12$.

A more limited set of data is available for Hawaiian basalts (table 1). The $[^{232}\text{Th}/^{238}\text{U}]$ data for these (Newman *et al.* 1984) provide an estimate of source $[\text{Th/U}] = 3.19 \pm 0.33$. Thus both the Iceland and Hawaii data have significantly higher source [Th/U] than do the normal ridges. Other ocean islands such as Reunion (Condomines *et al.* 1988) and Samoa (Newman *et al.* 1984) give even higher values for the source [Th/U], up to and even exceeding 4.0. Until more data are available it is not possible to obtain a reliable estimate of the average [Th/U] for the source of plume basalts.

There are more good quality Pb-isotope data for ocean island basalts than there are Th–U isotope data. It is possible to make estimates of mantle [Th/U] from Pb-isotopes ($\bar{\kappa}_{\text{Pb}}$) for Iceland, Hawaii and a number of other islands (table 2). $\bar{\kappa}_{\text{Pb}}$ values in table 2 range from 4.2 for Kerguelen and Tristan da Cunha to more typical values of 3.81 ± 0.03 for Iceland and 3.84 ± 0.04 for Hawaii to the lowest values so far of 3.75 ± 0.02 for the Australes.

4. Unsupported mantle Pb

Normal spreading ridges have a source [Th/U] which averages 2.58 ± 0.06. This ratio is significantly lower than the average time integrated [Th/U] ratio, $\bar{\kappa}_{Pb} = 3.8$, calculated from the radiogenic ^{208}Pb* and ^{206}Pb* abundances. The present-day mantle [Th/U] is too low to have produced this radiogenic Pb over 4.55 Ga and therefore the Pb is in effect 'unsupported' by the parent ^{232}Th and ^{238}U (Galer & O'Nions 1985).

In addition to Pb, radiogenic ^{87}Sr* is also unsupported by parent ^{87}Rb in the ridge basalt source (Tatsumoto et al. 1965). In the case of Rb–Sr this is because Rb is highly depleted in the upper mantle, having been removed to the continental crust, whereas Sr, being a more compatible element, is depleted to a relatively minor extent. However, it is less obvious that this situation should exist for Pb, because, like Th, U and Rb, it is a highly incompatible element and is also strongly concentrated into the continents. Galer & O'Nions (1985) suggested that Th, U and Pb are efficiently stripped from the upper mantle by melting, and have residence time there of less than 1 Ga and probably as short as 0.6 Ga. Their abundance must then be maintained by introduction of material from a less depleted region with [Th/U] ≈ 3.8 into the convecting mantle sampled by the ridges. The basis of their argument for a short residence time of Pb is reproduced in figure 1, and rests simply on the demonstration that there is a restriction on the length of time that mantle Pb could have been resident in region with [Th/U] $= 2.6$ and still have a value of $\bar{\kappa}_{Pb}$ of 3.80. In this simple model τ is the time at which [Th/U] is reduced to 2.6, the present upper mantle value, from a value of 3.9. This is a two-stage evolution model, the first stage being 4.55 Ga to τ, with [Th/U] $= 3.9$, and the second stage $\tau > t \geqslant 0$ with [Th/U] $= 2.6$. From figure 1 it is evident that $\{^{208}$Pb*$/^{206}$Pb*$\}$ is less than the values observed in the ridge basalts when τ increases beyond 0.6 to 1.0 Ga. The restriction imposed by $\{^{208}$Pb*$/^{206}$Pb*$\}$ is only removed if the first stage, 4.5 Ga to τ, takes place in a region with [Th/U] closer to 4.5 than 3.9. With one exception (Allègre et al. 1986) this value is higher than recent estimates for bulk Earth [Th/U] of 3.9 ± 0.1, 3.8 ± 0.2 and 3.9 ± 0.1 by Manhes et al. (1979), Goldstein & Galer (1991) and Rocholl & Jochum (1991) respectively, and is also higher than all measurements from plume basalts (table 2).

The short residence times for Pb, Th and U implied by these considerations require that these elements are stripped efficiently from the mantle by melting. This is certainly to be expected from recent work by McKenzie & O'Nions (1991) on partial melt distributions derived from the inversion of rare-earth element abundances. The melting processes involved are those that transport Pb, Th and U from the convecting region into the continental crust and mechanical boundary layer. Such transport can involve large melt fractions in plumes and extensional environments, but also small melt fractions in the source regions of island arc and alkali basalts.

The principal question now becomes the source from which Th, U and Pb are introduced into the ridge source. Two possibilities were considered by Galer & O'Nions (1985): the continental lithosphere, and mantle beneath the lower thermal boundary layer of upper mantle convection. The continental crust was considered to

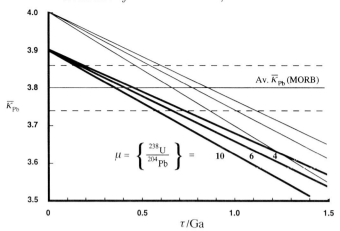

Figure 1. The evolution of Pb-isotopes in a reservoir in which $[^{232}\mathrm{Th}/^{238}\mathrm{U}] = 3.9 \equiv \kappa_1$ at 4.55 Ga, but is reduced to $2.6 \equiv \kappa_2$ at time τ in the past. Results are expressed in terms of $\bar{\kappa}_{\mathrm{Pb}}$ (mantle), the time-integrated $[^{232}\mathrm{Th}/^{238}\mathrm{U}]$ ratio for the whole interval 4.55 Ga to 0 required to produce the resultant radiogenic $^{208}\mathrm{Pb}^*/^{206}\mathrm{Pb}^*$. The dependence of $\bar{\kappa}_{\mathrm{Pb}}$ (mantle) on the time of $[\mathrm{Th}/\mathrm{U}]$ fractionation, τ, is shown for values of $\mu(= \{^{238}\mathrm{U}/^{204}\mathrm{Pb}\})$ of 4, 6 and 10, which encompass estimates for the MORB source. The heavy solid lines are constructed for $\kappa_1 = 3.9$, the plume basalt average (table 2), and the thin solid lines for $\kappa_1 = 4.0$, the estimated upper limit to the bulk Earth ratio of 3.9 ± 0.1 (Manhes *et al.* 1979). The horizontal line shows the ratio $\bar{\kappa}_{\mathrm{Pb}}$ (MORB), the average value for mid-ocean ridge basalts (table 2), and the dotted lines the $\pm\sigma$ range. The model $\bar{\kappa}_{\mathrm{Pb}}$ (mantle) values are less than $\bar{\kappa}_{\mathrm{Pb}}$ (MORB) for times of fractionation greater than *ca.* 600 Ma. Figure adapted from Galer & O'Nions (1985). Pb isotope evolution is modelled using

$$\{^{208}\mathrm{Pb}/^{204}\mathrm{Pb}\}_0 = \{^{208}\mathrm{Pb}/^{204}\mathrm{Pb}\}_\mathrm{I} + \kappa_1\mu_1(\exp(\lambda_{232}T) - \exp(\lambda_{232}\tau)) + \kappa_2\mu_2(\exp(\lambda_{232}\tau) - 1)$$

and

$$\{^{206}\mathrm{Pb}/^{204}\mathrm{Pb}\}_0 = \{^{206}\mathrm{Pb}/^{204}\mathrm{Pb}\}_\mathrm{I} + \mu_1(\exp(\lambda_{238}T) - \exp(\lambda_{238}\tau)) + \mu_2(\exp(\lambda_{238}\tau) - 1),$$

where $T = 4.55$ Ga, $\tau =$ time of Th/U fractionation, and λ_{232}, λ_{238} are decay constants for $^{232}\mathrm{Th}$ and $^{238}\mathrm{U}$. 'I' and 'o' are Pb ratios at 4.55 and 0 Ga, κ and μ are the $\{^{232}\mathrm{Th}/^{238}\mathrm{U}\}$ and $\{^{238}\mathrm{U}/^{204}\mathrm{Pb}\}$ ratios respectively, with subscripts 1 and 2 denoting the values for the intervals 4.55 Ga $< t < \tau$ and $\tau < t < 0$. $\kappa_1 = 3.9$ and $\mu_1 = 8.0$. $\{^{208}\mathrm{Pb}^*/^{206}\mathrm{Pb}^*\}$ and $\bar{\kappa}_{\mathrm{Pb}}$ are calculated using equations (30) and (31) in the text.

be an unlikely source because its Pb-isotope composition does not coincide with that of the ridge source, and entrainment from a deeper mantle region was therefore considered most probable. Certainly plume basalts, which are the potential carriers of Pb from this region, have Pb-isotope compositions that fully overlap those of ridge basalts.

5. Melt residues and mantle depletion

There are two ways in which the [Th/U] ratio in the upper mantle can be reduced. Either Th can be extracted in preference to U by magma generation (Galer & O'Nions 1985), or U from the crust can be returned to the upper mantle by subduction of altered oceanic crust. Since there is no doubt that both U and Th have been efficiently transferred to the continental crust and mechanical boundary layer by melt movement, we use the same melting model discussed above to show that Galer & O'Nions's proposal can account for the [Th/U] ratio of MORB. It is, however, difficult to show that U is not preferentially transported into the upper mantle by subduction, because so little is yet understood about transport in subduction zones.

To be consistent, we continue to use activities, and in addition to a_p and a_r in the melt, define A_p and A_r to be those in the solid residue. Like a_r, A_r is not a true activity. The evolution equations then are

$$dA_p/dt = -\alpha_p A_p \tag{32}$$

and
$$dA_r/dt = -\alpha_d A_r. \tag{33}$$

We first consider the relationship between K_p and K_d that is required to leave a residue with $[\mathrm{Th/U}] = 2.6$. The integrated $[^{238}\mathrm{U}/^{232}\mathrm{Th}]$ in the solid residue after extraction for infinite time is related to $G_s(\infty)$

$$G_s(\infty) = \left(\frac{\lambda_p}{\lambda_d}\right)\left[\frac{^{238}\mathrm{U}}{^{232}\mathrm{Th}}\right]_s = \int_0^\infty A_p(t')\,dt' \Big/ \int_0^\infty A_r(t')\,dt'. \tag{34}$$

Since $A_p(0) = K_p u_p(0)$, $A_r(0) = K_d u_r(0)$, and $A_r(\infty) = A_p(\infty) = 0$, equation (34) gives

$$G_s(\infty) = (K_p \alpha_d/K_d \alpha_p)\,a_p(0)/a_r(0). \tag{35}$$

The ratio in the solid when melting starts $G_s(0)$ is obtained using equation (12) for the total $^{238}\mathrm{U}$ and $^{232}\mathrm{Th}$ per unit volume

$$G_s(0) = \lambda_p \bar{c}_p/\lambda_d \bar{c}_d = (F_p/F_d)\,a_p(0)/a_r(0). \tag{36}$$

Combining equations (35) and (36) and substituting for α_d and α_p using equation (7) gives

$$G_s(\infty) = (K_p/K_d)\,G_s(0). \tag{37}$$

If $G_s(0)/G_s(\infty) = 3.9/2.6$, then $K_d = 0.67K_p$. This result is consistent with the general observation that $K_d < K_p$.

If the $[\mathrm{Th/U}]$ ratio in the upper mantle is less than that of the bulk Earth, there must be a corresponding reservoir in which $[\mathrm{Th/U}]$ exceeds that of the bulk Earth. Though there is no evidence for such an effect in the measured continental values of $[\mathrm{Th/U}]$ or of $\bar{\kappa}_{\mathrm{Pb}}$, it is likely to be small and difficult to detect. Highly incompatible elements such as Th, U, Pb, Rb and Cs have abundances in the continental crust that correspond to complete stripping of 30 40% of the mantle by mass. As the argument above shows, these mantle concentrations are not likely to be residual, but are maintained by entrainment (Galer & O'Nions 1985). Though the upper mantle concentrations of these elements are not known precisely, because of their short residence times (Galer & O'Nions 1985) they are unlikely to exceed 10% of the primitive mantle abundances. For continental crust with $[\mathrm{Th/U}] = 3.9$ and $[\mathrm{Th}] = 4.5$ p.p.m., and an upper mantle with $[\mathrm{Th/U}] = 2.6$ and $[\mathrm{Th}] = 0.007$ p.p.b., the combined $[\mathrm{Th/U}]$ for the crust and mantle is 3.8. This calculation shows that melt extraction can produce a large change in $[\mathrm{Th/U}]$ in the upper mantle, where Th and U are strongly depleted relative to the bulk Earth, whereas the corresponding change to $[\mathrm{Th/U}]$ in the continents can be small, because most of the U and Th from the upper mantle is now concentrated in the continental crust and mechanical boundary layer.

The same melting model can be used to estimate the efficiency of U and Th extraction required to leave a depleted residue with $[\mathrm{Th/U}] = 2.6$. We assume that the melting occurs within a layer of thickness h as the solid material upwells at velocity v. As before we assume that $K_p \ll 1$, and also for simplicity that $\phi \ll K_p$, or fractional melting. If the melting rate Γ is constant, and the total melt fraction by volume that has been extracted when the matrix reaches the top of the layer is ϕ_t, then

$$\Gamma = v\rho\phi_t/h. \tag{38}$$

The retention efficiency, E_p, is the ratio of ^{238}U in the residual layer of thickness h to its initial value before melt extraction

$$E_p = \int_0^h A_p(z)\,dz/K_p a_p(0)\,h, \tag{39}$$

where z is the height above the base of the layer. Since $z = vt$, the integral in equation (39) can be converted to an integral over time. Assuming $A_p(h) = 0$ gives

$$\int_0^h A_p(z)\,dz = v\int_0^\infty A_p(t)\,dt = \frac{vK_p a_p(0)}{\alpha_p}. \tag{40}$$

Substitution of equation (40) into (39) gives the retention efficiency

$$E_p = v/\alpha_p = \rho\phi v/F_p\,\Gamma. \tag{41}$$

Since we assume $K_p \gg \phi$, substitution for F_p from equation (8) reduces (41) to

$$E_p = K_p/\phi_t. \tag{42}$$

If the material entrained into the upper mantle has bulk Earth Th and U contents of 70 and 18 ppb respectively, and is then depleted to 10% of these amounts, the required value of E_p is 0.1, or $\phi_t - 10K_p$. For values of K_p between 10^{-3} and 10^{-4}, the total melt extracted ϕ_t must be in the range of 10^{-2} to 10^{-3}. This calculation is only concerned with the melt fraction that is involved in the transport of Th and U into the continental crust and mechanical boundary layer, where it remains geochemically isolated from the convecting upper mantle. Though the inversion of rare earth element concentrations (McKenzie & O'Nions 1991) shows that the total melt fractions beneath Iceland and Hawaii are much larger than 10^{-2}, and will therefore strip both Th and U from their source regions, the resulting aseismic ridges are subducted with the oceanic lithosphere. Therefore such melting has no long term effect on the [Th/U] ratio, or on the upper mantle concentrations of Th and U. The two processes that are likely to be important are island arc volcanism and the movement of alkali basalts and other small melt fractions into the continental crust and mechanical boundary layer. The inversion calculations (McKenzie & O'Nions 1991) show that the melt fractions in both environments are 10^{-2} or less, and can therefore produce the observed [Th/U] ratio. Except beneath Archaean shields, where the mechanical boundary layer is sufficiently thick to prevent extensive melting, plumes beneath continents will strip Th and U from their source regions. They will therefore have little effect on the [Th/U] ratio of the upper mantle, though they will reduce the concentrations of both elements.

6. Concluding remarks

Modelling of melting and melt extraction suggests that the (^{230}Th/^{232}Th) activity ratios in basalt melts from normal spreading ridges will not in general yield correct estimates for mantle [Th/U], because melt extraction occurs slowly relative to the decay rate of ^{230}Th. Only where melts are extracted rapidly relative to the ^{230}Th decay can the (^{230}Th/^{232}Th) ratio be used for this purpose. [^{232}Th/^{238}U] data for ridge basalts provides a better estimate, provided that the effects of alteration and low pressure crystallization are unimportant. These effects are minimized by selecting only primitive basalt glasses, and these provide a best estimate of [Th/U] = 2.58 ± 0.09 for their source. This ratio is much lower than the bulk Earth value and the time-integrated [^{232}Th/^{238}U] ratios derived from the radiogenic daughter Pb in ridge basalts, as noted previously (Galer & O'Nions 1985). The view that the residence times of Th, U and Pb in the upper mantle must be relatively short, and probably less than 1 Ga, is confirmed.

The Th, U and Pb in the upper mantle are present at about 0.1 of their bulk Earth abundances, and must be sustained by entrainment from material outside of the convecting upper mantle region sampled by melting at ridges, probably by plumes. The observed [^{232}Th/^{238}U] shows that Th is extracted from the upper mantle in preference to U. Since aseismic ridges are subducted by trenches, plume melting in oceanic regions has no effect on the upper mantle concentration of U and Th. The extraction processes concerned are those that transfer these elements to the continental lithosphere, by generating melt beneath island arcs and continental interiors. Inversion calculations show that both often involve melt fractions of 1 % or less, and can therefore account for the observed [Th/U] of about 2.6.

We thank J. Gill and D. Macdougall for helpful comments, and the Natural Environmental Research Council and the Royal Society for their support.

References

Allègre, C. J. & Condomines, M. 1982 Basalt genesis and mantle structure studied through Th-isotope geochemistry. *Nature, Lond.* **299**, 21–24.

Allègre, C. J., Dupré, B. & Lewin, E. 1986 Thorium/uranium ratio of the Earth. *Chem. Geol.* **56**, 219–227.

Condomines, M., Hemond, Ch. & Allègre, C. J. 1988 U–Th–Ra radioactive disequilibria and magmatic processes. *Earth planet. Sci. Lett.* **90**, 243–263.

Cortini, M. 1984 Uranium in mantle processes. In *Uranium geochemistry, mineralogy, geology, exploration and resources* (ed. B. De Vivo, F. Ippolito, G. Capaldi & P. R. Simpson). Institution of Mining and Metalurgy.

Galer, S. J. G. & O'Nions, R. K. 1985 Residence time of thorium, uranium and lead in the mantle with implications for mantle convection. *Nature, Lond.* **316**, 778–782.

Goldstein, S. J., Murrel, M. T., Janecky, D. R., Delaney, J. R. & Clague, D. A. 1991 Geochronology and petrogenesis of MORB from the Juan da Fuca and Gorda ridges by ^{238}U–^{230}Th disequilibrium. *Earth planet. Sci. Lett.* **107**, 25–41.

Goldstein, S. L. & Galer, S. J. G. 1991 Limits on the Th/U ratio of continental crust and the Earth. *Terra* **3**, 486 (abstr.).

Jochum, K. P., Hofmann, A. W., Ito, E., Seufert, H. M. & White, W. 1983 K, U and Th in mid-ocean ridge basalt glasses and heat production K/U and K/Rb in the mantle. *Nature, Lond.* **306**, 431–450.

Langmuir, C. H., Bender, J. F., Bence, A. E., Hanson, G. N. & Taylor, S. R. 1977 Petrogenesis of basalts from the FAMOUS area: mid-atlantic ridge. *Earth planet. Sci. Lett.* **36**, 133.

Manhes, G., Allègre, C. J., Dupré, B. & Hamelin, B. 1979 Lead–lead systemics, the 'age of the Earth' and chemical evolution of our planet in a new representation space. *Earth planet. Sci. Lett.* **44**, 91–104.

McKenzie, D. 1985 ^{230}Th–^{238}U disequilibria and melting processes beneath ridge axes. *Earth planet. Sci. Lett.* **72**, 149–157.

McKenzie, D. & O'Nions, R. K. 1991 Partial melt distributions from inversion of rare-earth element concentrations. *J. Petrol.* **32**, 1021–1091.

Newman, S., Finkel, R. C. & Macdougall, J. D. 1984 Comparison of ^{230}Th–^{238}U disequilibrium systematics in lavas from three hot spot regions: Hawaii, Prince Edward and Samoa. *Geochim. cosmochim. Acta* **48**, 315–324.

Othman, D. B. & Allègre, C. J. 1990 U–Pb isotopic systematics at 13° N East Pacific Ridge segment. *Earth planet. Sci. Lett.* **98**, 120–137.

Rocholl, A. & Jochum, K. P. 1991 Th/U of the Earth: a chondritic perspective. *Terra* **3**, 486 (abstr.).

Rubin, K. H. & Macdougall, J. D. 1988 ^{226}Ra excesses in mid-ocean ridge basalts and mantle melting. *Nature, Lond.* **335**, 158–161.

Sigmarsson, O., Condomines, M. & Fourcade, S. 1992 Mantle and crustal contribution in the genesis of Recent basalts from off-rift zones in Iceland: constraints from Th, Sr and O isotopes. Preprint.

Tatsumoto, M., Hedge, C. E. & Engel, A. E. J. 1965 Potassium, rubidium, strontium, thorium, uranium and the ratio strontium-87 to strontium-86 in oceanic tholeiitic basalt. *Science, Wash.* **150**, 886–888.

Tatsumoto, M., Knight, R. J. & Allègre, C. J. 1973 Time differences in the formation of meteorites as determined from the ratio of lead-207 to lead-206. *Science, Wash.* **180**, 1279–1283.

Williams, R. W. & Gill, J. B. 1989 Effects of partial melting on the uranium decay series. *Geochim. cosmochim. Acta* **53**, 1607–1620.

Melts and metasomatic fluids: evidence from U-series disequilibria and Th isotopes

By J. B. Gill

Earth Sciences Department, University of California, Santa Cruz, California 95064, U.S.A.

Radioactive disequilibria and Th isotopes in volcanic rocks change in response to the extent and rate of partial melting, the attainment of chemical and isotopic equilibrium, and the presence and composition of a vapour phase before and during melting. Enrichments of ^{230}Th with respect to ^{238}U by 10–60% in MORB and intraplate alkalic magmas are attributed to the greater incompatibility of Th in small degree silicate partial melts. Inverse correlations between $(^{230}$Th$)/(^{238}$U$)$ and $(^{230}$Th$)/(^{232}$Th$)$ ratios indicate either that partial melting lasts more than 10^5 years, or that smaller degree melts take scores of millenia longer to move from the site of last chemical equilibrium with the source to the eruption site. Disequilibrium enrichments of ^{226}Ra with respect to ^{230}Th and of ^{231}Pa with respect to ^{235}U by 150–300% occur in oceanic island arc basalts and MORB. The excess ^{226}Ra in arcs is related to magma genesis, but the other disequilibria may reflect alteration or assimilation within the crust. When these large disequilibria accompany magma genesis they indicate transfer times between source and surface of a few decades to a few millenia and suggest an open system or chemical disequilibrium during melting. Th and U isotope and concentration systematics for lamproites suggest that melting was fast enough to preclude chemical equilibrium between source and melt. Non-silicate fluids may fractionate U-series radionuclides differently and to a greater extent than do silicate melts. This property may explain the U–Th–Ra characteristics of carbonate- and water-rich melts, and distinctive Th isotopic compositions in magmas derived from metasomes in the subcontinental lithosphere.

1. Introduction

Natural radioactive disequilibria provide unique information about the chemistry and timescale of magma genesis. Secular radioactive equilibrium is when the numbers of atoms (N) of all nuclides in a decay series are inversely proportional to their decay constants (λ), so that their 'activities' ($\lambda * N$) are equal. Equilibrium is reached after about five times the half-life of the longest-lived intermediate nuclide. Disequilibrium is created when two nuclides in the same decay series are fractionated, so that their activities become unequal. The basic principles are illustrated in figure 1. Although it refers specifically to the ^{238}U–^{230}Th pair, it can be generalized to all parent–daughter pairs in the three decay schemes. Recent reviews of the principles involved, and of the application of disequilibria studies to other aspects of magma genesis such as differentiation, are by Condomines *et al.* (1988), Gill *et al.* (1992), and Gill & Condomines (1992). This paper expands on one section of the latter reference.

Conventions used in this paper are as follows. Parentheses around a nuclide denote activities in dpm g^{-1}, whereas brackets denote concentrations in p.p.m. The

Phil. Trans. R. Soc. Lond. A (1993) **342**, 79–90

Printed in Great Britain

79

© 1993 The Royal Society

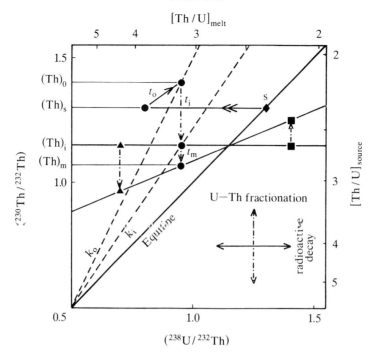

Figure 1. Schematic (^{230}Th)–(^{238}U) isochron diagram. Although it refers specifically to the ^{238}U–^{230}Th pair, it can be generalized to all parent–daughter pairs in the three decay schemes. Disequilibrium is created by fractionating ^{230}Th from ^{238}U during partial melting. Magma is produced by melting source S which is isotopically homogeneous at the scale of melting and initially in radioactive equilibrium (that is, on the Equiline) with $(\text{Th})_s$. If $D_{\text{Th}} < D_{\text{U}}$, then $(^{238}\text{U})/(^{230}\text{Th})$ or $k < 1.0$ in the melt. As the degree of melting increases, the ^{230}Th-enrichment and $[\text{Th}/\text{U}]$ ratio in the melt decrease. In the example illustrated, the degree of melting is insufficient for the $[\text{Th}/\text{U}]_{\text{melt}}$ to decrease to the $[\text{Th}/\text{U}]_{\text{source}}$. At the end of melting (after time t_o), the $(^{230}\text{Th})/(^{232}\text{Th})$ ratio in the melt is $(\text{Th})_o$, which exceeds $(\text{Th})_s$ if t_o is long. The $[\text{Th}/\text{U}]_{\text{melt}}$ remains constant after the melt ceases to be in chemical equilibrium with its source, but the $(^{230}\text{Th})/(^{232}\text{Th})$ ratio decreases during transfer time t_i to $(\text{Th})_i$ at the time of eruption as the result of unsupported radioactive decay. The ^{230}Th-enrichment decreases from k_o at the end of melting to k_i at the time of eruption. Near the time of eruption the melt precipitates phenocrysts, for some of which $D_{\text{Th}} < D_{\text{Th}}$ (illustrated by the solid triangle) and for others of which $D_{\text{Th}} < D_{\text{U}}$ (illustrated by the solid square). There is isotopic homogeneity, all phases having $(\text{Th})_i$. During the time between eruption and measurement (t_m), all phases evolve toward radioactive equilibrium by pivoting around $(\text{Th})_i$ on the Equiline such that the bulk sample has $(\text{Th})_m$ when measured. In this case, t_m is the conventional age of the lava.

$(^{238}\text{U})/(^{230}\text{Th})$ activity ratio is referred to as k, and the $(^{226}\text{Ra})/(^{230}\text{Th})$ activity ratio as l. The $(^{230}\text{Th})/(^{232}\text{Th})$ activity ratio of a volcanic rock at the time of measurement is referred to as $(\text{Th})_m$; at the time of eruption it is $(\text{Th})_i$, at the end of melting (i.e. at the last chemical equilibration with the source) it is $(\text{Th})_o$, and at the beginning of melting the ratio in the source is $(\text{Th})_s$ (i.e. the initial isotopic composition of the source) (figure 1). The corresponding times between events during the history of a volcanic rock are: t_m, the time between eruption and measurement (the conventional age of the rock); t_i, the time between the end of melting and eruption (the transfer time from the source); and t_o, the mean time between the beginning and end of melting (the duration of partial melting).

The t_o interval is ambiguous for two reasons. First, if dynamic melting occurs throughout a 50 km high column of decompressing mantle, t_0 is almost zero where the column first crosses its solidus, but can be a million years at the top of the column; for this paper, t_0 is the mean age for any specific aggregate of melts. Second, t_0 and t_i are separated by the time at which chemical equilibrium ceases between melt and source. At extremes, these intervals may correspond to the time of permeable flow of primitive melt through the asthenosphere versus ascent and differentiation in sheets or pipes through the lithosphere, respectively. However, a continuum in the porosity and spacing of melt channels is likely. Chemical equilibrium will be reached when channels are closely spaced and solid state diffusion is rapid; chemical disequilibrium is enhanced by the opposite and is likely even during permeable flow in the asthenosphere (Spiegelman & Kenyon 1992).

Some advantages of using short-lived radionuclides to study melting are that disequilibria provide quantitative measurements of trace element fractionation during magma genesis, disequilibria constrain the temporal and perhaps spatial scale of magma formation, Th responds quickly to chemical modifications of the source and is the only isotopic tracer which reflects a current trace element ratio, and Th and Ra are the most incompatible isotopic tracers in silicate melts.

2. Equilibrium melting

(a) Background

For a quarter century, basalts have been related to different degrees of batch partial melting of mantle peridotite under chemical and isotopic equilibrium (Green & Ringwood 1967; Gast 1968; Minster & Allegre 1978; Hofmann & Hart 1978). The degree of melting was tacitly assumed to equal a melt fraction or porosity (ϕ) that could be calculated by, for example, inverting trace element concentrations of basalts using equations which assume chemical equilibrium between all the melt and all the solid matrix. Effects of melting rate and melt path were ignored. Many trace elements and primitive samples are needed in order to allow inversion of the data because both the Ds (partition coefficients) and the initial source composition must be known to calculate ϕ. Tholeiitic magmas, such as those erupted at mid-ocean ridges or Kilauea volcano, Hawaii, were found to represent relatively large degrees of melting (10–20%) of fertile peridotite, with more alkalic magmas reflecting smaller percent melts. There were taken to be the ϕs necessary to produce interconnectivity of the melt.

The need to know the source composition and to use undifferentiated samples can be relaxed if there is radioactive equilibrium at the onset of melting and differentiation does not alter activity ratios. Only the Ds of parent and daughter nuclides are necessary in order to calculate ϕ from the observed disequilibrium.

(b) ^{230}Th–^{238}U *disequilibria and* $(^{230}Th)/(^{232}Th)$ *ratios*

Both Th and U are highly incompatible elements and the partition coefficients of Th and U are thought to be similar and very low such that the [Th/U] ratio of melt would differ from that of source peridotite only in highly alkalic magmas. Discovery of 18% excess ^{230}Th ($k = 0.85$) in MORB on average (Condomines *et al.* 1981; Newman *et al.* 1983; Goldstein *et al.* 1990, 1991) was, therefore, surprising. Excess ^{230}Th has now been found in dozens of samples in more than ten laboratories by both mass and alpha spectrometry.

If related to melting, it implies that the mantle was permeable at low porosity and

$D_{Th} < D_U$, so that Th was preferentially enriched at small degrees of melting (McKenzie 1985). Continuous models that involve fractional melting above a small ϕ, and dynamic models involving mixing of small percent melts from variously depleted levels of an ascending source, are required to reconcile the low k values with other geochemical results. If small percent melts or melts from more fertile sources, mix with larger percent melts or melts from more depleted sources, the former will dominate the trace element characteristics of the mixture, including radioactive disequilibria (Williams & Gill 1989; Goldstein *et al.* 1991).

[Th/U] and [Th] decrease with increasing degrees of partial melting until $[Th/U]_{melt}$ equals $[Th/U]_{source}$. The values of D_U and D_{Th} determine how much total melting is required to achieve this equality; for example, if $D_U = 0.015$ and $D_{Th} = 0.005$, then about 15% melting is required (Williams & Gill 1989). Even if there is chemical and isotopic equilibrium during melting, $(Th)_i$ may differ from $(Th)_s$, and be higher in larger degree melts, for either of two reasons. First, greater degrees of partial melting will be accompanied by higher $(Th)_o$ if t_o is longer than about one half-life of ^{230}Th (Williams & Gill 1989). The amount of increase is a function of partition coefficients and porosity as well as melting rate and time, but the effect is expected to be less than 15% when upwelling rates are greater than 5–10 cm a^{-1}, depending on the maximum depth of melt extraction. If disequilibria are integrated into geochemical models which constrain some of the variables independently for a specific location, then melting rates and geometry can be deduced. On the other hand, $(Th)_i$ will be less than $(Th)_o$, and lower in smaller melt fractions, if melts take millenia to move to the surface from the site of last chemical equilibrium with the source. This effect will be enhanced if smaller melt fractions have longer transfer times than larger melt fractions. Which of the two effects predominates depends on whether chemical equilibrium is maintained until final melt extraction. If it is maintained, $(^{230}Th)/(^{232}Th)$ ratios will rise because the matrix is U-enriched; if chemical equilibrium is not maintained, they fall because the melt is Th-enriched. This is why the change in convention between $(Th)_o$ and $(Th)_i$ occurs at the time of last chemical equilibration with the source.

Whether t_o or t_i is long enough to cause $(Th)_i$ to differ from $(Th)_s$, or whether the total degree of melting is large enough for $[Th/U]_{melt}$ to equal $[Th/U]_{source}$, depends on melting models and rates, time, and the partition coefficients of Th and U. No study of disequilibria has yet been sufficiently integrated with other trace element and isotopic measures of source homogeneity and percent of melting to address this topic rigorously. However, discovery that k ratios correlate inversely with Th contents and [Th/U] ratios at similar Sr and Th isotope ratios between the Endeavor and southern Juan de Fuca spreading segments in the NE Pacific is an important first step (Goldstein *et al.* 1990, 1991). These results are consistent with chemical and isotopic equilibrium and homogeneity at the scale of melting beneath the two segments, and indicate smaller percent melting beneath the Endeavor. Following the logic above, the 6% higher $(Th)_i$ in the southern Juan de Fuca segment could result from: (i) long t_o (*ca.* 300 Ka to reach 15% melting using the parameters of Williams & Gill (1989)) followed by rapid ascent; or (ii) instantaneous t_o followed by about 22–40 Ka longer t_i for the smaller melt fraction depending on assumptions about $[Th/U]_{source}$. Case (i) assumes $(Th)_o = 1.3$ (i.e. $[Th/U]_{source} = 2.3$) whereas case (ii) assumes $(Th)_o = 1.5$ or higher. Case (i) is preferable because it better fits the global correlation of Th and Sr isotopes and case (ii) is inconsistent with excess ^{226}Ra if it was acquired during partial melting (see below).

Most oceanic and continental alkali basalts have more ^{230}Th enrichment than ocean island tholeiites (see summaries by Condomines *et al.* 1988; Gill *et al.* 1992), as expected if disequilibrium is primarily a function of the degree of melting. However, no global correlation exists between the amount of excess ^{230}Th and major or trace element measures of percent melting. In part this is because the latter cannot distinguish the effects of smaller percent melting from greater source enrichment. It is also because the extent of radioactive disequilibrium includes the effects of melting rate, partition coefficients, and approach to chemical and isotopic equilibrium (see below) which vary from place to place.

Permeable flow of small degree melts has chromatographic effects, including separation of elements according to small differences in their partition coefficients (Navon & Stolper 1987; McKenzie & O'Nions 1991). This can result in nonlinear mixing effects amongst the isotopic tracers which will be most noticeable for Th because it is the most incompatible tracer except for He. These effects will contribute to variations in disequilibria and to scatter in isotope correlation diagrams involving Th.

(c) $(^{226}$Ra$)/(^{230}$Th$)$ and $(^{231}$Pa$)/(^{235}$U$)$ disequilibria

Discovery of other, even larger disequilibria followed. Some MORB and island arc volcanic rocks have $(^{226}$Ra$)/(^{230}$Th$)$ and $(^{231}$Pa$)/(^{235}$U$)$ ratios of 1.5 to 3.0 (Capaldi *et al.* 1983; Rubin & Macdougall 1988; Reinitz & Turekian 1989; Rubin *et al.* 1989; Gill & Williams 1990; Williams & Perrin 1989; Williams *et al.* 1991). Most of the excess ^{226}Ra in arc lavas is attributed to melting processes because it is present in primitive magmas and it decreases in more differentiated lavas of individual volcanoes.

The results for MORB currently are more ambiguous and require further study of well-characterized samples for clarification. Excess ^{226}Ra appears to be a primary magmatic feature, unrelated to surface alteration (Reinitz & Turekian 1989), although the large enrichments of ^{226}Ra in submarine hydrothermal fluids makes alteration a constant concern. Magmatic assimilation of young hydrothermally altered wallrocks or sediments is another possible source of excess ^{226}Ra (Sigmarsson *et al.* 1991). If, however, Ra–Th and Pa–U fractionations occur during partial melting, then some important implications follow.

$(^{226}$Ra$)$–$(^{230}$Th$)$ disequilibria can be used to estimate $t_0 + t_i$ if $D_{Ra} = D_{Ba}$, the [Ba/Th] ratio of a source is constant, and the $(^{226}$Ra$)/(^{230}$Th$)$ ratios of magmas at the end of partial melting can be estimated from their Ba/Th ratios. Data for some MORB suggest that this sum is about 1 ka (Rubin & Macdougall 1990). Two problems with this approach should be noted, however, in addition to the obvious assumption of no chemical fractionation between Ra and its homologue Ba. First, even at constant D_{Ra} and D_{Th}, $(^{226}$Ra$)/(^{230}$Th$)$ ratios vary inversely with melting rate (Williams & Gill 1989). Second, there is inherent ambiguity about the Ba/Th ratio of the source. The 1 ka time estimate above assumed a Ba/Th ratio of 8 for the source of MORB from the East Pacific Rise. This is an order of magnitude lower than usually adopted, and probably indicates considerable source heterogeneity.

The above approach cannot be extended to disequilibrium between ^{231}Pa and ^{235}U because no homologue of Pa is known. None the less, the enrichments of ^{231}Pa and ^{230}Th correlate positively even though the ^{231}Pa enrichments are considerably larger (Williams *et al.* 1991; Goldstein *et al.* 1991). They are of the same magnitude as ^{226}Ra enrichments. If the excess ^{230}Th is related to melting, then the excess ^{231}Pa probably is also, and $D_{Pa} < D_{Th} < D_{U}$. These results also suggest local consistency in the extent of ^{231}Pa enrichment. That is, if one assumes that $(^{231}$Pa$)/(^{235}$U$)$ ratios are constant in the neovolcanic zone and attributes changes in this ratio to time since eruption, then the calculated age (t_m) for off-axis samples is the same as obtained from ^{238}U–^{230}Th disequilibria by assuming constant (Th)$_i$.

3. Disequilibrium melting and accessory minerals

'Disequilibrium melting' has at least three quite different meanings. *Radioactive disequilibrium* is the most likely and accompanies melting whenever parent and daughter elements partition differently into the melt, which is common (see above). *Chemical disequilibrium* occurs whenever melting is too rapid for chemical potentials to be the same in melt and matrix, such that element concentrations in the melt do not reflect the bulk partition coefficients of the residue. The rate-limiting process is volume diffusion in the matrix, which is thought to require 10^4–10^6 years in cm-diameter mantle pyroxenes and garnet for highly charged cations (Bédard 1989). If t_0 is shorter, as implied above, then chemical disequilibrium is likely. The highly incompatible character of Th in melts combined with its slow diffusion in solids results in a high Peclet number (Speigelman & Kenyon 1992) which makes Th the least likely isotopic tracer to achieve chemical equilibrium during permeable flow. *Isotopic disequilibrium* is the least likely and occurs only when the source is isotopically heterogeneous at its solidus and the isotopic composition of the melt depends on the melting mode (O'Nions & Pankhurst 1974). However, Th is more likely to be isotopically heterogeneous in the subsolidus than Sr, because Th responds faster to metasomatism and diffuses more slowly. Consequently, ^{238}U and ^{230}Th may approach radioactive equilibrium within phases faster than Th isotopes are homogenized between phases by diffusion.

Accessory minerals in magma sources can play an important role in all three kinds of disequilibria if they have high concentrations and highly fractionated parent/daughter ratios. Several accessory minerals which may dissolve in or coexist with small degree melts in the mantle have quite high [Th/U] ratios (e.g. phlogopite, amphibole, whitlockite); others have quite low [Th/U] ratios (carbonates, titanates, and perhaps garnet). Any of these would contribute greatly to radioactive disequilibria if they are refractory or if chemical equilibrium is not achieved. Melts in chemical equilibrium with a residual accessory phase acquire trace element characteristics which are the inverse of that phase, in proportion to its weight fraction in the residue (figure 2a). This effect is extended during dynamic melting, where melts of more fertile sources routinely blend with melts of more depleted sources. Trace elements can be affected by accessory phases which are residual only in the more fertile source regions whereas major elements are determined by the average extent of melting of the entire melt column (Klein & Langmuir 1987; Williams & Gill 1989; Reagan & Gill 1989). In contrast, an accessory phase consumed during static melting is irrelevant; its elements are simply repartitioned between refractory phases and larger degree melts (figure 2b).

Chemical disequilibrium during melting can produce melts with the trace element characteristics of the accessory phase itself when that phase dominates the melting mode. The concentrations of trace elements in the melt remain constant despite variations in the degree of melting as long as the melting mode is constant (Bédard 1989). If the accessory phase is old enough for all phases to be isotopically

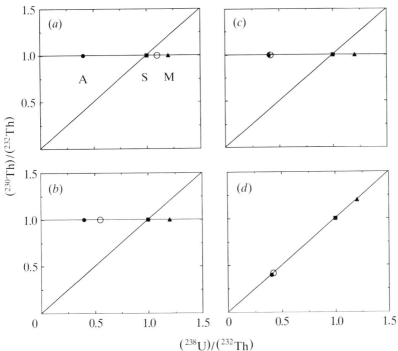

Figure 2. Schematic $^{238}U–^{230}Th$ isochron diagrams illustrating the effects of various melting models. In each, the solid circle 'A' represents a subsolidus accessory phase with a high Th/U ratio and $D_{Th} \gg D_U$ (e.g. phlogopite), the solid triangle 'M' represents the matrix (the rest of the source) having bulk $D_{Th} < D_U$, the solid square 'S' represents the bulk source in radioactive equilibrium at the onset of melting, the open circle represents the liquid formed during partial melting, and both t_o and t_i are small so that $(Th)_i = (Th)_s$. (a) There is chemical and isotopic equilibrium during melting (the simplest case). The accessory phase is refractory and its partition coefficients are sufficiently large to cause the liquid to be slightly ^{238}U-enriched. (b) As in (a) except that the accessory phase is consumed during melting so that the $^{238}U/^{232}Th$ ratio is only a function of the percent melting and the bulk partition coefficients of the matrix. (c) There is Th isotopic equilibrium but not chemical equilibrium during melting. The $^{238}U/^{232}Th$ ratio of the melt approaches that of the accessory phase because it dominates the melting mode. (d) As in (c) except that there is isotopic as well as chemical disequilibrium during melting. If the accessory phase reaches secular equilibrium before Th isotopes are homogenized by diffusion, then chemical disequilibrium during melting results in radioactive equilibrium such that the isotopic composition of Th is that of the accessory phase instead of the bulk source.

homogeneous despite differences in parent/daughter ratio, then the melt inherits the radioactive disequilibrium of the accessory phase (figure 2c).

If there is isotopic as well as chemical disequilibrium during melting, almost anything is possible. Melts will become enriched in the faster diffusing nuclide if diffusion coefficients differ substantially. Ironically, the melt can even wind up in radioactive equilibrium if the time between metasomatism and melting is long with respect to the half-lives involved but too short for all phases to be homogenized isotopically via diffusion (figure 2d). Therefore, chemical disequilibrium during melting can lead to radioactive disequilibrium (figure 2c), but isotopic disequilibrium during melting can lead to radioactive equilibrium (figure 2d)!

All three kinds of disequilibrium may be especially common during the initial melting of a thermal boundary layer in which accessory metasomatic minerals were precipitated during the freezing of earlier fluids and melted during later thermal perturbations. Two examples are the base of the subcontinental lithosphere and the base of the mantle wedge immediately above subducting oceanic crust. These two cases can differ in both the metasomatic fluids involved (hence the kind of radioactive disequilibrium produced in accessory phases when isotopic equilibrium is achieved), and in the time between metasomatism and melting (hence the extent to which isotopic equilibrium is achieved). An example of each is discussed below.

Metasomatism in the subcontinental lithosphere can predate magma genesis by many millions of years and produce minerals with extreme Th/U ratios that entirely melt near the solidus. One example is the role of phlogopite in lamproite genesis. A recent study of Th–U disequilibria in the 56 ka Gaussberg lamproites suggests that these magmas formed too quickly for chemical equilibrium to be achieved. This conclusion rests on the high precision for $(Th)_m$ and [Th/U] ratios achieved using mass spectrometry (Williams *et al* 1992) Th and Sr isotopes correlate negatively within the lavas, indicating an isotopically heterogeneous source. However, Th, U, and HFSE concentrations and ratios remain constant in the lavas. The association of greater ^{230}Th excess with higher Sr contents is attributed to smaller degrees of melting. If the isotopic heterogeneity of the lavas reflects variable proportions of old metasomatic phlogopite in the source, then there is a direct correlation between the earlier amount of metasomatism and the later degree of melting. These features could be produced by consumption of all phlogopite during very small degrees of partial melting under chemical equilibrium (figure 2b). However, the large amounts of excess ^{230}Th and the high Th concentrations are more easily explained if chemical equilibrium was incomplete. In neither case was there chemical equilibrium with residual phlogopite, so that small degree melts of the subcontinental lithosphere acquired the trace element, isotopic, and radioactive disequilibria characteristics of the accessory phase (phlogopite) which dominated the melting mode.

In contrast, metasomatism of the mantle wedge immediately above subducting lithosphere may precede partial melting beneath the volcanic arc by only a few hundred millenia at most. This is the time necessary for convection of the mantle from the maximum pressure stability of amphibole in the slab (i.e. the depth at which maximum wedge metasomatism occurs) to the maximum pressure stability of amphibole in peridotite (i.e. the depth at or above which melting initiates) (see Tatsumi 1989; Davies & Stevenson 1992; Gill *et al.* 1993). This is sufficient time to achieve radioactive equilibrium between ^{230}Th and ^{238}U, but not homogeneous Th isotope ratios between newly formed amphibole (or other accessory metasomatic minerals) and pre-existing phases. As in the Gaussberg example, lack of chemical equilibrium can produce melts with the [Th/U] and $(^{230}Th)/(^{232}Th)$ ratios of the accessory phase. However, both the accessory phase and the melt will be in radioactive equilibrium in the subduction zone example (figure 2d). The difference reflects the time between metasomatism and melting, and may explain the more widespread ^{230}Th–^{238}U equilibrium in arcs compared to other tectonic environments (M. Condomines, personal communication).

4. Fluids and mantle metasomatism

Discovery of extreme Ra and U enrichment in carbonatite called attention to the fractionation of radionuclides by non-silicate fluids (Williams *et al.* 1986; Pyle *et al.* 1991). Carbonate- and halogen-rich C–O–H fluids are present in the mantle, able to flow through it because of their low dihedral angles, and likely to leave both patent and cryptic metasomatism in their wake (Watson *et al.* 1990; Wallace & Green 1988).

The strong dependence of the water-solubility of U on oxidation state at low pressure is well known (Langmuir 1978). Recent experiments have also shown the importance of both Cl and CO_3^{2-} contents in aqueous fluids for the complexing and solubility at high pressure of trace elements including U and Ba (Keppler & Wyllie 1991; Brennan & Watson 1991). Consequently, the igneous geochemistry of Ra, Pa, U, and Th is likely to be sensitive to and informative about the present and past distribution of such fluids.

One obvious application is to the magnitude of the ^{226}Ra–^{230}Th and ^{231}Pa–^{235}U disequilibria observed in some MORB and arc lavas. Eighteen percent excess ^{230}Th in MORB, although surprising, is at least compatible with currently accepted values of 0.5–2% for ϕ during mantle melting if $(D_U - D_{Th})$ is as large as 0.005 (Williams & Gill 1989). Qualitatively, excess Ra with respect to Th is consistent with peridotite-basalt Ds if Ra^{2+} behaves like Ba^{2+}. Quantitatively, however, melting models that assume partitioning of elements between silicate solids and liquid cannot account for the extent of radioactive disequilibria observed. For example, if $l = 3.0$, then $\phi = \frac{1}{2}D_{Th}$ (McKenzie 1987); the required ϕ is absurdly low because D_{Th} is nearly zero.

It was noted earlier that these large disequilibria may prove to be related to alteration or assimilation. However, if they are created during partial melting, the most promising explanations infer the entrainment of Ra and Pa from larger source regions than Th and U (i.e. open systems) (Cortini 1984; Condomines *et al.* 1988; Gill & Williams 1990), their preferential concentration during permeable flow (Navon & Stolper 1987), or chemical disequilibrium during melting. The first requires either the addition of Ra ± Pa-bearing fluids to the source, causing flux-induced melting at the vapour-present solidus, or the extraction of subsolidus vapour adjacent to a partly molten region. In either case, a free vapour phase is necessary. The second postulates that permeable flow of melt through the mantle is analogous to solution flow through a chromatographic column. Chromatography assumes chemical equilibrium and maximizes small differences in D. Chemical disequilibrium relies on differences in diffusion coefficients and the effects of accessory phases as discussed above. The lability of short-lived nuclides in radiation damaged sites, and recoil effects from surfaces to melts would also contribute to the effects of chemical disequilibrium.

In general, Ra enrichment is greater in island arc lavas and perhaps mid-ocean ridge basalts than in tholeiitic or alkalic intraplate basalts, where it is usually less than 30% and often zero. This does not seem to reflect longer t_i because low Ra enrichments characterize mafic magmas from volcanoes with high levels of historical activity (e.g. Kilauea, Reunion, Nyamuragira, Mt Cameroon). If related to melting processes, the distribution of Ra enrichment implies that the open system or permeable flow conditions characteristic of active plate margins (both divergent and convergent) are absent within plates. This, in turn, suggests either that subsolidus fluid is absent at sites of intraplate melting or that fluid flow regimes differ in the absence of suction at 'corners' such that more of the flow path is characterized by chemical disequilibrium (Spiegelman & McKenzie 1987; Ribe 1987).

The most dramatic example of fluid-caused disequilibria in igneous rocks also may be the least applicable; the extreme Ra and U enrichments in the Na-carbonatites of Oldoinyo Lengaii which has erupted twice in the past 35 years (Williams *et al.* 1986; Pyle *et al.* 1991). Both times it contained excess ^{228}Ra as well as ^{226}Ra in a proportion that correlates with repose time and suggests formation of new carbonatite shortly after each eruption ceases. The most likely genetic mechanism is exsolution from a silicate magma at low pressure. Although this is unlikely to be common during the genesis of common magmas, it is analogous to the vapour exsolution which may be widespread during mantle metasomatism (Wallace & Green 1988).

A second volcano at which carbonate-rich fluid has been important during magma genesis is Nyiragongo, Zaire (Vanlerberghe *et al.* 1987; Williams & Gill 1992; C. Deniel & A. Demant, personal communication). In general, ^{238}U-enriched lavas erupted first, contain groundmass calcite, and have high $(Th)_i$ and the lowest $^{87}Sr/^{86}Sr$ ratios of the Virunga field. Younger (including historical) lavas are ^{230}Th-enriched or in equilibrium and closer to the Sr–Th mantle array. The variations are consistent with a decreasing CO_2 flux through the volcano and with decreasing influence of a carbonate-metasomatized source with time.

Hydrous rather than carbonate fluids are implicated by disequilibria in Gaussberg lamproites (see above). $(Th)_i$ for the Gaussberg lavas also lie above the Sr–Th mantle array, but the divergence at Gaussberg is toward higher rather than lower $^{87}Sr/^{86}Sr$ ratios, in contrast to the results for Nyiragongo. The negative correlation between Th and Sr isotope ratios is attributed to variable amounts of phlogopite in the source. The phlogopite accounts for the increase in [Th/U] ratio during metasomatism such that the modern [Th/U] ratio of the source indicated by Th isotopes (κ_{Th}) exceeds the time-integrated [Th/U] ratio of the source indicated by Pb isotopes (κ_{Pb}) (Williams *et al.* 1992). This pattern is the opposite of that observed in oceanic basalts, indicates that the subcontinental lithosphere includes Th-enriched reservoirs complementary to the depleted mantle, and is consistent with the [Th/U] of the Earth being about 4.2 (Williams & Gill 1992)

I am indebted to R. Williams, M. Condomines, and M. Reagan for discussion, none of whom agrees with all that is said here, and to the Royal Society for making a manuscript prerequisite for accepting their invitation to this meeting.

References

Bédard, J. H. 1989 Disequilibrium mantle melting. *Earth planet. Sci. Lett.* **91**, 359–366.

Brennan, J. M. & Watson, E. B. 1991 Partitioning of trace elements between olivine and aqueous fluids at high *P–T* conditions; implications for the effect of fluid composition on trace element transport. *Earth planet. Sci. Lett.* **107**, 672–688.

Capaldi, G., Cortini, M. & Pece, R. 1983 U and Th decay-series disequilibria in historical lavas from the Eolian islands, Tyrrhenian Sea. *Isotope Geosci.* **1**, 39–55.

Condomines, M., Bouchez, R., Ma, J. L., Tanguy, J. C., Amosse, J. & Piboule, M. 1987 Short-lived radioactive disequilibria and magma dynamics in Etna volcano. *Nature, Lond.* **325**, 607–609.

Condomines, M., Hemond, C. H. & Allègre, C. J. 1988 U–Th–Ra radioactive disequilibria and magmatic processes. *Earth planet. Sci. Lett.* **90**, 243–262.

Condomines, M., Morand, P. & Allègre, C. J. 1981 ^{230}Th–^{238}U radioactive disequilibria in tholeiites from the FAMOUS zone: Th and Sr isotopic geochemistry. *Earth planet Sci. Lett.* **55**, 247–256.

Cortini, M. 1984 Uranium in mantle processes. In *Uranium geochemistry, mineralogy, geology, exploration and resources* (ed. B. DeVivo, F. Ippolito, G. Capaldi & P. R. Simpson), pp. 4–11. London: Institute of Minerals and Metals.

Davies, J. H. & Stevenson, D. J. 1992 Physical model of source region of subduction zone volcanics. *J. geophys. Res.* **97**, 2037–2070.

Gast, P. W. 1968 Trace element fractionation and the origin of tholeiitic and alkaline magma types. *Geochim. cosmochim. Acta* **32**, 1057–1086.

Gill, J. & Condomines, M. 1992 Short-lived radioactivity and magma genesis. *Science, Wash.* **257**, 1368–1376.

Gill, J., Morris, J. & Johnson, R. 1993 Timescale for producing the geochemical signature of island arc magmas: U–Th–Po and B–Be systematics in recent Papua New Guinea lavas. *Geochim. cosmochim. Acta.* (Submitted.)

Gill, J., Pyle, D. & Williams, R. 1992 Igneous rocks. In *Uranium series disequilibrium: applications to environmental problems*, 2nd edn, ch. 7, 207–258 (ed. M. Ivanovich & R. Harmon). Oxford: Clarendon Press.

Gill, J. & Williams, R. 1990 Th isotope and U-series studies of subduction-related volcanic rocks. *Geochim. cosmochim. Acta* **54**, 1427–1442.

Green, D. H. & Ringwood, A. E. 1967 The genesis of basaltic magmas. *Contrib. Miner. Petrol.* **15**, 103–190.

Goldstein, S. J., Murrell, M. T. & Janecky, D. R. 1990 Th and U isotopic systematics of basalts from the Juan de Fuca and Gorda ridges by mass spectrometry. *Earth planet. Sci. Lett.* **96**, 134–146.

Goldstein, S. J., Murrell, M. T., Janecky, D. R., Delaney, J. R. & Clague, D. A. 1991 Geochronology and petrogenesis of MORB from the Juan de Fuca and Gorda ridges by ^{238}U–^{230}Th disequilibrium. *Earth planet. Sci. Lett.* **109**, 255–272.

Hofmann, A. W. & Hart, S. R. 1978 An assessment of local and regional isotopic equilibrium in the mantle. *Earth planet. Sci. Lett.* **38**, 44–62.

Keppler, H. & Wyllie, P. J. 1991 Partitioning of Cu, Sn, Mo, W, U, and Th between melt and aqueous fluid in the systems haplogranite–H_2O–HCl and haplogranite-H_2O–HF. *Contrib. Mineral Petrol.* **109**, 139–150.

Klein, E. M. & Langmuir, C. H. 1987 Global correlations of ocean ridge basalt chemistry with axial depth and crustal thickness. *J. geophys. Res.* **92**, 8089–8115.

Langmuir, D. 1978 U solution-mineral equilibrium at low temperatures with applications to sedimentary ore deposits. *Geochim. cosmochim. Acta* **42**, 547–569.

McKenzie, D. 1985 ^{230}Th–^{238}U disequilibrium and the melting process beneath ridge axes, *Earth planet. Sci. Lett.* **72**, 149–157.

McKenzie, D. 1987 The compaction of igneous and sedimentary rocks. *J. geol. Soc. Lond.* **144**, 299–307.

McKenzie, D. & O'Nions, R. K. 1991 Partial melt distributions from inversion of rare earth element concentrations. *J. Petrology* **32**, 1021–1091.

Minster, J. F. & Allègre, C. J. 1978 Systematic use of trace elements in igneous processes. III. Inverse problem of batch partial melting in volcanic suites. *Contrib. Miner. Petrol.* **69**, 37–52.

Navon, O. & Stolper, E. 1987 Geochemical consequences of melt percolation: the upper mantle as a chromatographic column. *J. Geology* **95**, 285–307.

Newman, S., Finkel, R. C. & Macdougall, J. D. 1983 ^{230}Th–^{238}U disequilibrium systematics in basalts from 21 N on the East Pacific Rise. *Earth planet Sci. Lett.* **65**, 17–23.

O'Nions, R. K. & Pankhurst, R. J. 1974 Petrogenetic significance of isotope and trace element variations in volcanic rocks from the Mid-Atlantic. *J. Petrology* **15**, 603–634.

Pyle, D., Dawson, J. B. & Ivanovich, M. 1991 Short-lived decay series disequilibria in the natrocarbonatite lavas of Ol Doinyo Lengai, Tanzania: constraints on the timing of magma genesis. *Earth planet. Sci. Lett.* **105**, 378–386.

Reagan, M. K. & Gill, J. 1989 Coexisting calcalkaline and high-Nb basalts from Turrialba volcano, Costa Rica: implications for residual titanates in arc magma sources. *J. geophys. Res.* **94**, 4619–4633.

Reinitz, I. & Turekian, K. 1989 ^{230}Th/^{238}U and ^{226}Ra/^{230}Th fractionation in young basaltic glasses from the East Pacific Rise. *Earth planet. Sci. Lett.* **94**, 199–207.

Ribe, N. M. 1987 Theory of melt segregation – a review. *J. Volc. geothermal Res.* **33**, 241–253.

Rubin, K. H. & Macdougall, J. D. 1988 ^{226}Ra excesses in mid-ocean ridge basalts and mantle melting. *Nature, Lond.* **335**, 158–161.

Rubin, K. H. & Macdougall, J. D. 1990 Dating of neovolcanic MORB using (^{226}Ra/^{230}Th) disequilibrium. *Earth planet. Sci. Lett.* **101**, 313–322.

Rubin, K. H., Wheller, G. E., Tanzer, M. O., Macdougall, J. D., Varne, R. & Finkel, R. 1989 [238]U decay series systematics of young lavas from Batur volcano, Sunda arc. *J. Volc. geothermal Res.* **38**, 215–226.

Sigmarsson, O., Cohen, A. S. & O'Nions, R. K. 1991 U–Th–Ra systematics in hydrothermal systems at spreading ridges. *Eos, Wash.* **72**, 574.

Spiegelman, M. & Kenyon, P. 1992 The requirements for chemical disequilibrium during magma migration. *Earth planet. Sci. Lett.* **109**, 611–620.

Spiegelman, M. & McKenzie, D. P. 1987 Simple 2D models for melt extraction at mid-ocean ridges and island arcs. *Earth planet. Sci. Lett.* **83**, 137–152.

Tatsumi, Y. 1989 Migration of fluid phases and genesis of basalt magmas in subduction zones. *J. geophys. Res.* **94**, 4697–4707.

Vanlerberghe, L., Hertogen, J. & Macdougall, J. 1987 Geochemical evolution and Th–U isotope systematics of alkaline lavas from Nyiragongo volcano. *Terra Cognita* **7**, 367.

Watson, B., Brennan, J. M. & Baker, D. R. 1990 Distribution of fluids in the continental mantle. In *Continental mantle* (ed. M. A. Menzies), pp. 111–126. Oxford: Clarendon Press

Wallace, M. E. & Green, D. H. 1988 An experimental determination of primary carbonatite magma composition. *Nature, Lond.* **355**, 343–346.

Williams, R. W., Collerson, K., Gill, J. B. & Deniel, C. 1992 High Th/U ratios in subcontinental lithospheric mantle: mass spectrometric measurement of Th isotopes in Gaussberg lamproites. *Earth planet. Sci. Lett.* **111**, 257–268.

Williams, R. W. & Gill, J. B. 1989 Effects of partial melting on the Uranium decay series, *Geochim. cosmochim. Acta* **53**, 1607–1619.

Williams, R. W. & Gill, J. B. 1992 Th isotopes and U-series disequilibria in some alkali basalts. *Geophys. Res. Lett.* **19**, 139–142.

Williams, R. W., Gill, J. B. & Bruland, K. W. 1986 Ra–Th disequilibria systematics: timescale of carbonatite magma formation at Oldoinyo Lengai volcano, Tanzania. *Geochim. cosmochim. Acta* **50**, 1249–1259.

Williams, R. W., Murrell, M. T. & Goldstein, S. J. 1991 Large [231]Pa enrichments in neovolcanic MORB measured by mass spectrometry. *Eos, Wash.* **72**, 295.

Williams, R. W. & Perrin, R. E. 1989 Measurement of [231]Pa–[235]U disequilibrium in young volcanic rocks by mass spectrometry. *Eos, Wash.* **70**, 1398.

Mantle heterogeneity beneath oceanic islands: some inferences from isotopes

By Mark D. Kurz

Chemistry Department, Woods Hole Oceanographic Institution, Woods Hole, Massachusetts 02543, U.S.A.

Radiogenic isotopes in oceanic basalts are extremely useful as tracers of long-lived heterogeneities in the Earth's mantle. Helium isotopes provide unique information in that high ^{3}He/^{4}He ratios are indicative of relatively undegassed mantle reservoirs (i.e. mantle with high time-integrated ^{3}He/(Th+U) ratios). An alternative hypothesis is that high ^{3}He/^{4}He ratios may have been produced by ancient melting events, if the solid/melt partition coefficient (K_d) for He is greater than that for Th and U (i.e. yielding relatively high He/(Th+U) in the residue of melting). However, the distribution of helium within basaltic phenocrysts, and olivine/glass helium partitioning within mid-ocean ridge basalts, suggest that helium behaves as an incompatible element during melting (K_d (olivine/glass) < 0.0055), which strongly supports the hypothesis that high ^{3}He/^{4}He ratios are derived from undegassed mantle reservoirs.

Isotopic measurements of He, Sr, and Pb in Hawaiian volcanoes lavas demonstrate that the mantle sources have changed on extremely short timescales, between 100 and 10000 years before present. The preferred explanation for these variations is that they represent heterogeneities within the Hawaiian mantle plume, combined with late stage melting in the lithosphere for post shield alkali basalts. Helium isotopic data from Kilauea, Hualalai and Mauna Loa suggest that the plume is presently located beneath Kilauea (and Loihi seamount), and constrain the melting zone of the Hawaiian plume to be less than 40 km in radius.

1. Introduction

Basalts erupted at oceanic islands provide important constraints on the nature and origin of heterogeneity within the Earth's mantle. Isotopic variations between oceanic islands demonstrate that the mantle is heterogeneous on various length scales, and that the heterogeneities have persisted for at least 10^9 years (Hofmann & Hart 1978). However, it is not possible to infer accurately the geometry of mantle isotopic heterogeneities because basaltic lavas do not retain information about the original depth of their sources; at best, one can infer the depth of melt segregation. Because geochemical heterogeneities have persisted in the mantle on 10^9 year timescales, many geochemists favour layered mantle convection models (e.g. Allegre & Turcotte 1985). However, whole mantle convection models can also allow such long-lived heterogeneities, if for example the lower mantle is not well stirred (Davies 1984; Gurnis & Davies 1986). Although the geochemical data are important constraints, they alone cannot distinguish between these two distinct models for mantle structure (Silver *et al.* 1988).

Phil. Trans. R. Soc. Lond. A (1993) **342**, 91–103

Printed in Great Britain

Helium isotopic data are important because high ^{3}He/^{4}He ratios in basalts are generated by high ^{3}He/(Th + U) ratios in the mantle, and thus must be derived from the least degassed mantle. However, some recent laboratory partitioning experiments indicate that the helium solid/melt partition coefficients could be higher than for Th and U. If so, high ^{3}He/^{4}He ratios could be produced by early melting events in the mantle, which would leave the residue of melting with high He/(Th + U) ratios. High ^{3}He/^{4}He ratios cannot be produced by recycled materials, such as subducted oceanic crust sediments or ancient continental lithosphere, which have low ^{3}He/(Th + U) ratios and hence low ^{3}He/^{4}He ratios (Mamyrin & Tolstikhin 1984). The simplest way to explain a relatively undegassed terrestrial reservoir is to assume that it has remained far from the surface. Indeed, most studies of ^{3}He flux from the mantle require some source within the lower mantle (O'Nions & Oxburgh 1983; Kellogg & Wasserburg 1990). Many oceanic islands, such as Hawaii, Iceland, Reunion, and Samoa have ^{3}He/^{4}He ratios higher than MORB, and based on helium alone, these islands would be derived from the least degassed mantle sources, which could reside in the lower mantle. The helium solid/melt partitioning is critical to these inferences.

Several processes may complicate such simple interpretations of isotopic data. Recent fluid dynamical studies have emphasized the importance of entrainment processes within the mantle (Griffiths & Campbell 1990; Griffiths & Campbell 1991). Entrainment of surrounding mantle can result in significant heterogeneity within individual mantle plumes. In addition, there is evidence that small melt fractions are rapidly removed from the mantle, suggesting that many basaltic liquids are produced by fractional melting (McKenzie 1985; Richter 1986; Riley & Kohlstedt 1991). Therefore, the basaltic liquids that are extruded on the Earth's surface may be hybrids of many different melt fractions. Interaction between the rising melt and the mantle may also result in geochemical changes resulting from 'percolation' effects (Navon & Stolper 1986; McKenzie & O'Nions 1991). If the geochemistry of erupted volcanic rocks is to be effectively used as a 'window into the mantle', criteria must be developed to distinguish between effects caused by melting processes and those related to source characteristics. The hypothesis advanced here is that one important way to constrain this problem is with data from individual volcanoes, where geological and temporal constraints are available. The emphasis here is on the isotopes of helium, because of the unique constraints high ^{3}He/^{4}He ratios can place on mantle derived rocks, and because of the large helium isotopic variability that is found within individual volcanoes.

2. Helium partitioning

Although there have been few studies of crystal-melt noble gas partitioning, most workers in this field have tended to assume that helium is an incompatible element within mantle minerals (Kurz *et al.* 1982*a*; McKenzie & O'Nions 1991). However, several laboratory equilibration studies have yielded noble gas crystal/melt partition coefficients significantly higher than Th and U and other incompatible trace elements (Hiyagon & Ozima 1986; Broadhurst *et al.* 1992). Hiyagon *et al.* (1986) found helium olivine/melt partition coefficients close to 0.1, with higher values for the heavy noble gases. If helium is a mildly compatible element ($K_d \approx 0.1$), then melting could conceivably leave the residue of melting with higher ^{3}He/(Th + U) ratio, because the

Table 1. *Olivine/glass partitioning for* MORB *glass sample ALV 526-1*

(This sample is an olivine rich glass (*ca.* 30% olivine), and approximately 1% vesicles; further petrographic and geological details can be found in Bryan & Moore (1977).)

glass ^{4}He concentration[a]	3.117×10^{-6} cc (STP) g^{-1}
olivine ^{4}He concentration[a]	2.875×10^{-8} cc (STP) g^{-1}
olivine ^{4}He within inclusions[b]	1.136×10^{-8} cc (STP) g^{-1}

$$K_d(\text{ol/melt}) \ll \frac{(^4\text{He}_{\text{olv}} - {}^4\text{He}_{\text{olv,crush}})}{^4\text{He}_{\text{glass}}}$$

$$K_d(\text{ol/melt}) \ll 0.0058$$

[a] Total concentrations for olivine and glass were obtained by melting of 0.5 to 2 mm grains in an ultra-high vacuum furnace.
[b] Helium contained by inclusions is inferred to be that released by crushing, although this is a minimum because some helium may also be dissolved within the glassy inclusions (see text).

Th and U crystal/melt partition coefficients are very low ($\lesssim 10^{-3}$ (La Tourrette *et al.* 1992)). As a result, mantle reservoirs that were melted early in Earth's history could have high ^{3}He/^{4}He ratios, which would invalidate the assumption that high ^{3}He/^{4}He ratios indicate undegassed mantle sources.

There is some evidence that helium is an incompatible element during partial melting. The glassy rims of mid-ocean ridge basalts typically have helium concentrations of 10^{-5}–10^{-6} cc(STP) g^{-1} (Kurz & Jenkins 1981; Kurz *et al.* 1982*b*; Sarda & Graham 1990). Basaltic phenocrysts have significantly lower helium concentrations, typically 10^{-8} cc(STP) g^{-1} (Kurz *et al.* 1982*a*), which would imply bulk partition coefficients less than 0.01. Unfortunately, large olivine phenocrysts are rare in MORB, and most of the phenocryst concentration measurements have been performed on ocean island basalts, where there are few other gas containing phases.

Table 1 lists new helium measurements on coexisting olivine and glass from the FAMOUS area at 37° on the mid Atlantic ridge. The concentration data provided by this experiment suggest that the olivine/glass partition coefficient (K_d) is significantly less than 0.0058. This value is an upper limit for several reasons. First, even submarine glasses are known to have undergone some degassing before, or during, eruption on the sea floor (Kurz & Jenkins 1981; Sarda & Graham 1990); hence the measured helium concentration in the glass must be considered to represent a *lower* limit estimate for the magma at eruption. On the other hand, the concentration in the olivine is an *upper* limit due to the presence of melt inclusions within olivine. Even though the calculation of the partition coefficient (see table 1) includes a correction for the helium released by crushing, there could still be considerable amounts of helium trapped within the glass inclusions that would not be released by crushing in vacuum.

There is also evidence, from the distribution of helium within olivine phenocrysts, that helium is incompatible in the olivine matrix. Most of the helium contained by olivine is released by crushing *in vacuo* (Kurz *et al.* 1982*a*), which implies that the helium resides within melt or fluid inclusions. This is illustrated by figure 1, which shows the helium distribution within olivine crystals from a number of Mauna Loa (Hawaii) lava flows. These data were obtained by crushing the samples in vacuo, followed by melting of the powder (Kurz *et al.* 1990). In all cases, most of the helium is released by crushing, as indicated by ratios of ^{4}He$_{\text{olv}}$/^{4}He$_{\text{tot}}$ always significantly less than unity. The data define a trend whereby higher total concentrations are

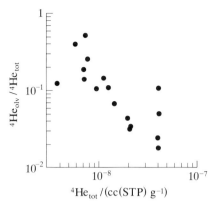

Figure 1. Partitioning within olivine crystals from Mauna Loa Volcano (Kurz *et al.* 1987, 1990). The total concentrations reflect the sum of helium extracted by crushing and melting of 1–2 mm olivines *in vacuo*. The ratio of $^4\mathrm{He_{olv}}/^4\mathrm{He_{tot}}$ is the fraction that is contained by melt inclusions; $^4\mathrm{He_{olv}}$ is the amount of helium obtained by melting the previously crushed olivine powder, and $^4\mathrm{He_{tot}}$ is the sum of crushing and melting the olivines. This fraction decreases with increasing total concentration, and is inferred to be related to abundance of melt inclusions (see text). Note that all of the samples have $^4\mathrm{He_{olv}}/^4\mathrm{He_{tot}}$ significantly less than one, which suggests that helium is not contained significantly by the olivine matrix, but is held within the inclusions. The variability in the $^4\mathrm{He_{olv}}/^4\mathrm{He_{tot}}$ ratio at high total $^4\mathrm{He}$ concentration is primarily related to incomplete crushing of several samples.

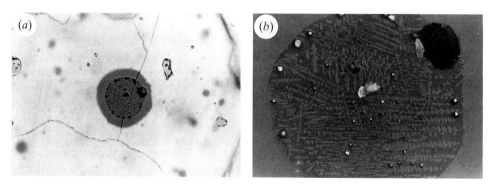

Figure 2. Photographs of an olivine grain containing a large melt inclusion exposed on a polished surface, from the 1868 lava flow of Mauna Loa, in transmitted and reflected light. (*a*) The spherical melt inclusion in the centre of the field is approximately 200 µm (width of field 1.2 mm, magnification 50 ×). The lighter area surrounding the inclusion is the olivine host crystal; small dark patches are smaller melt and spinel inclusions within the olivine. (*b*) This higher magnification photograph (field of view *ca.* 250 µm, magnification 250 ×) shows that the melt inclusion has undergone post-entrapment crystallization, as evidenced by skeletal pyroxene crystals. The large round dark area at the top of the field is a bubble, as are the smaller ones dispersed throughout the inclusion. The angular dark area near the centre of the inclusion is an ion microprobe crater.

accompanied by greater fraction released by crushing (i.e. lower $^4\mathrm{He_{olv}}/^4\mathrm{He_{tot}}$). This suggests that high helium concentrations are associated with inclusions (which release helium upon crushing).

Microscopic examination of the olivine crystals supports the hypothesis that melt inclusions control the helium concentrations in olivine. Figure 2 shows photographs of one crystal containing melt inclusions, with an inclusion exposed on the polished surface. The photograph demonstrates that the melt inclusion contains microscopic bubbles which are the most likely site for helium and other gases. These bubbles could be produced by post-entrapment crystallization (as suggested by the spherulitic crystal growth in the glass inclusion), or may exist before entrapment. The existence of a gas phase in these melt inclusions is probably related to confinement pressure on eruption, and speed of quenching, so they may not be present in all olivine crystals. Microscopic examination of the samples shown in figure 2 also suggests that fluid inclusions (i.e. containing liquid CO_2) are rare. Therefore, the features shown in figure 2, combined with the data in figure 1 strongly suggest that melt inclusions dominate the helium abundance in olivine phenocrysts, which is consistent with the interpretation that helium is an incompatible element in olivine.

Although the data presented here only provide an upper limit K_d for olivine/melt, they do illustrate that helium partition coefficients inferred from natural crystals are significantly lower than those found by laboratory experiments (Hiyagon & Ozima 1986). One explanation is that the laboratory experiments were also affected by the presence of microscopic inclusions, as discussed by Hiyagon & Ozima. Additional data will be necessary to resolve this discrepancy. Because all existing data from naturally formed olivines suggests that helium behaves as an incompatible element, it is assumed that high $^3He/^4He$ ratios relate to degassing history.

3. Global isotopic patterns

Isotopic variations in basalts are generally interpreted to reflect mixing between different reservoirs; various efforts to explain the isotopic variability in basalts have suggested the necessity of four or five distinct mantle sources (White 1985; Zindler & Hart 1986a; Hart 1988; Hart et al. 1992). Figure 3 summarizes most of the existing coupled He, Sr and Pb isotopic data for oceanic islands. Hawaii, Iceland and Samoa have the highest $^3He/^4He$ ratios, and by the reasoning adopted here, their lavas are derived from the least degassed mantle sources. The tholeiites from Loihi seamount, which have the highest $^3He/^4He$ ratios, have Sr and Nd isotopic compositions which differ significantly from those expected for bulk Earth. In fact, Hawaii is a geochemical end-member only with respect to helium; as illustrated in figure 3, data from Hawaiian basalts fall within the centre of the Sr-Pb diagram. Those oceanic islands having $^3He/^4He$ ratios lower than MORB values (i.e. lower than *ca.* $7 R/R_{atm}$), such as Tristan da Cunha, Gough and Sao Miguel, may be derived from mantle sources having significant contributions from recycled oceanic crust or sediments (Kurz et al. 1982a).

One problem that is well illustrated by figures 3 and 4 is that there are significant local variations in many of the oceanic islands. At Hawaii for example, two volcanoes Loihi and Mauna Loa, which are only 40 km apart, define most of the global variability in $^3He/^4He$ ratios, ranging from 8 to 32 times atmospheric. One explanation for this variation in helium, accompanied by small variability in Sr and Pb, is that helium is decoupled from the other isotopes by accumulation of ^4He during residence in a magma chamber (Condomines et al. 1983; Zindler & Hart 1986b), or by metasomatism in the mantle (Vance et al. 1989). However, in most cases where He, Sr, Nd and Pb are available for the same samples, there are correlations between them. This includes Haleakala and Mauna Loa on Hawaii (Kurz et al. 1987; Kurz & Kammer 1991), Samoa (Farley et al. 1992), Sao Miguel (Kurz et al. 1992). In Reunion basalts none of the isotopic systems, including helium, show any significant variability (Staudacher et al. 1990; Graham et al. 1990). There are few general correlations between any isotope systems in lavas from Iceland and

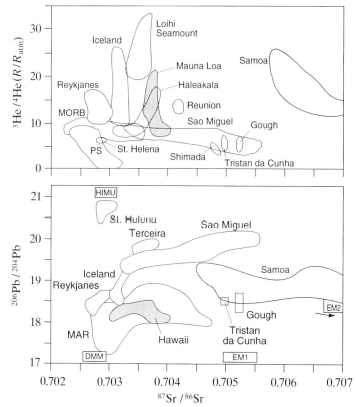

Figure 3. Helium, lead, and strontium isotopic data summary for selected oceanic islands (Kurz *et al.* 1982*a*, 1983, 1992; Farley *et al.* 1990; Graham *et al.* 1990; Poreda *et al.* 1986; and references in the text). PS refers to Pacific seamounts (Graham *et al.* 1988). Note that in the Sr–Pb diagram, Hawaiian basalts display a limited variation compared to other oceanic islands. The labels for distinct mantle reservoirs (HIMU), and enriched mantle (EM), proposed by Zindler & Hart (1986*a*) are shown for references.

Loihi seamount (Condomines *et al.* 1983; Staudigel *et al.* 1984), but this incoherence may relate to the lack of age information for most of the samples. Therefore, the systematic relationships between He, Sr and Pb in oceanic island basalts, particularly those from Hawaii, argue strongly against decoupling of helium from the other isotopes by magma chamber production of ^{4}He, or by metasomatism, and these processes will not be considered in any detail here (see Kurz & Kammer (1991) for discussion).

Because high ^{3}He/^{4}He ratios indicate undegassed mantle sources, helium is central to understanding the origin of mantle heterogeneity. Advocates of whole mantle convection have suggested that ^{3}He is coming from the core (Davies 1990), and others have used high ^{3}He/^{4}He ratios as indicators of a lower mantle origin for volcanism (Allegre *et al.* 1983; Hart *et al.* 1992). In evaluating such hypotheses, it is critical to evaluate the relationships between helium and the isotopes of Sr, Nd, and Pb, particularly with respect to small-scale heterogeneities, in order to discern possible mixing end-members.

4. The local context: inferences from Hawaiian volcanoes

There are many oceanic islands with large isotopic variations over small time or length scales, and a proper understanding of the causes of these variations is of fundamental importance. Because basalts from Hawaii have the highest ^{3}He/^{4}He ratios, understanding the local variations shown in figure 3 is particularly important to noble gas work. As one of the largest, longest-lived and presently most active hotspots, Hawaii, provides an important record of melting in the mantle. The fact that Hawaii constitutes the highest known mass flux at a hotspot (Sleep 1990) may suggest a relationship between high ^{3}He/^{4}He and material flux from the lower mantle. This relationship does not necessarily extend to other islands, because many of the islands having ^{3}He/^{4}He higher than MORB have low calculated mass fluxes (for example Iceland, Bouvet, and Reunion (Sleep 1990)). However, it is worth noting that Hawaii is the oceanic island that has both the highest mass flux and ^{3}He/^{4}He ratios.

At Hawaiian volcanoes it is well established that shield tholeiites are isotopically distinct from the post-erosional alkali basalts (Chen & Frey 1983, 1985). The Hawaiian shield building tholeiites have systematically higher ^{3}He/^{4}He than younger alkali basalts (Kurz *et al.* 1983; Kurz *et al.* 1987). These data are consistent with the mantle plume model for Hawaiian volcanism because they suggest an evolutionary trend, whereby the shield building stages are produced by melting of plume material with higher ^{3}He/^{4}He (Kurz *et al.* 1983; Kaneoka 1983; Kurz *et al.* 1987; Garcia *et al.* 1990). The later stages then are presumed to have ^{3}He/^{4}He similar to MORB due to a melting of MORB related upper mantle. The mechanism by which the later stages have different isotopic signatures, often presumed to result from melting of the lithosphere, could involve melt percolation because all the Hawaiian extrusives must pass through a substantial thickness of lithosphere (50–100 km).

An important aspect of the Hawaiian isotopic variations is the short timescale over which they occur at Mauna Loa volcano (Kurz *et al.* 1987; Kurz & Kammer 1991). Kurz & Kammer showed that the isotopic composition of Sr, Pb and He in Mauna Loa basalts decreased significantly 10000 years ago, and there was another change in Sr and Pb isotopes approximately 600 years ago, with ^{3}He/^{4}He remaining essentially constant. The timescale is provided by geological mapping of historical and radiocarbon dated lava flows (Rubin *et al.* 1987). This complex isotopic evolution is demonstrated in figure 4*a* along with He and Sr data from some other Hawaiian volcanoes. The data from Mauna Loa provides the important time information, and demonstrates that alkali basalts and tholeiites, even from the same volcanic edifice, may not necessarily be connected by mixing processes, unless the sampling is on a short enough timescale.

There are two alternative explanations for the isotopic variations shown in figure 4*a*. Kurz & Kammer suggested changes in the source materials, as Mauna Loa migrates away from the hotspot. The isotopic end-members required by this model are shown sequentially in figure 4*b*, and include the plume, another distinct source, perhaps from a different part of the plume, and finally the lithosphere. Heterogeneity within the plume could be produced by entrainment processes (Griffiths & Campbell 1991). The lithospheric involvement becomes important only after the volcano is removed from the plume melting conduits, and some of the heat from the plume melts the lithosphere (Liu & Chase 1991).

McKenzie & O'Nions (1991) suggested that the isotopic variations within Mauna

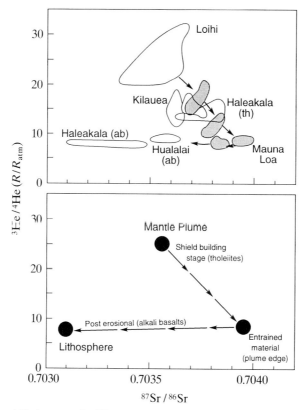

Figure 4. (*a*) Sr and He isotopes for Hawaiian basalts with the assumed temporal evolution pattern denoted by the direction of the arrows. The Mauna Loa basalts (shaded) are critical to this evolution scheme because the 'bend' in the evolution trend is observed between 600 and 2000 years before present (Kurz *et al.* 1983, 1987; Staudigel 1984; Kurz & Kammer 1991; and unpublished data, this laboratory). (*b*) Inferred isotopic compositions for the different components in the mantle beneath Hawaii that are required to explain the variability in the top figure. The highest ^{3}He/^{4}He ratios are inferred to be related to the mantle plume itself. The decrease in ^{3}He/^{4}He combined with increased ^{87}Sr/^{86}Sr is assumed to be related to heterogeneity within the plume, perhaps by entrainment affects that could lead to such axial variations (Griffiths & Campbell 1991). The lowest ^{3}He/^{4}He and ^{87}Sr/^{86}Sr ratios, found in the alkali basalts, are inferred to be related to melting in the lithosphere.

Loa (figure 4*a*) could be explained by mixing between two mantle sources via a melt percolation model. In their model, helium is removed from the plume first due to its extreme incompatibility within silicates, while the other elements such as strontium are released into the melt more slowly. The predicted time evolution is not unlike that shown in figure 4*a*. In the context of the Hawaiian isotope data, there are several important problems with the McKenzie & O'Nions model. The model predicts that radiogenic isotopes and incompatible trace element abundances will decrease with time at any volcanic shield. As pointed out by Frey & Rhodes (this symposium), this prediction is at odds with existing Rb/Sr, ^{87}Sr/^{86}Sr, Sm/Nd, and ^{143}Nd/^{144}Nd data for Kilauea and Mauna Loa. In addition, preliminary rare earth data for the Mauna Loa samples shown in figure 4*a*, does not display a systematic variation with time, as would be predicted by the model (Kurz & Shimizu, unpublished ion microprobe data for melt inclusions).

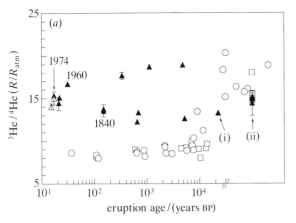

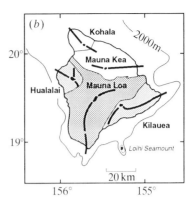

Figure 5. Helium isotopic evolution for three Hawaiian volcanoes, Hualalai (□), Kilauea (▲), and Mauna Loa (○). The timescale is provided by historical flows, radiocarbon dates on some lava flows (Rubin *et al.* 1987), and stratigraphy. There is no timescale beyond 30000 years because this is the age limit of the radiocarbon dates (Kurz *et al.* 1983, 1987, 1990; Kurz & Kammer 1991; and unpublished data this laboratory). Hollow triangles are volcanic gases from Kilauea (Craig & Lupton 1976; Jenkins *et al.* 1977). (i) Hilina series; (ii) dredged tholeiites.

Mauna Loa is the largest volcano on the island of Hawaii (and on Earth): an important question is whether the basalts erupted from the two smaller adjacent active volcanoes, Hualalai and Kilauea, have displayed similar isotopic variations. Figure 5 shows the temporal helium isotopic record from these two volcanoes (Kurz *et al.* 1990). As discussed earlier, the main feature of the Mauna Loa temporal record is the marked decrease in $^3He/^4He$ roughly 10000 years ago. For the Hualalai basalts, the temporal helium isotopic evolution is similar to Mauna Loa; the dredged tholeiites have higher $^3He/^4He$ than the historical alkali basalts (note that all surficial Hualalai eruptions are alkali basalts). As mentioned above, this is consistent with previous data from Haleakala (Kurz *et al.* 1987) and with the idea that the alkali basalts are derived from melting of MORB related lithosphere/asthenosphere, because the $^3He/^4He$ converge to MORB values (roughly $8\,R/R_{atm}$ in figure 5).

During the past 10000 years, basalts from Kilauea have consistently higher $^3He/^4He$ ratios than the other two volcanoes (between 12 and $20\,R/R_{atm}$). The consistently higher $^3He/^4He$ ratios suggest that Kilauea eruptions are more directly associated with the mantle plume, and therefore that the mantle plume contributes material (i.e. ^3He) to Kilauea, but not to Hualalai or Mauna Loa. This places a limitation on the size of the plume at approximately 40 km (see also Frey & Rhodes, this symposium). In addition, the data from Kilauea are significantly more variable, with helium isotopic differences between the 1840 and 1960 flows implying variability on the 100 year timescale. During the past 10000 years Mauna Loa and Hualalai basalts have had $^3He/^4He$ ratios which are indistinguishable from MORB values (*ca.* $8\,R/R_{atm}$). There are several possible causes for the short-term helium isotopic variation in Kilauea basalts. Based on trace element abundance variations, Rhodes *et al.* (1989) proposed that there may be periodic invasion of the Kilauea plumbing systems by Mauna Loa magma. Because the helium isotopic composition of the two volcanoes is so distinct, this could yield a significant change in the isotopic composition. This is feasible because Mauna Loa magma batches are so much larger than Kilauea; Klein (1982) estimated that the Mauna Loa magma reservoir is

4-2

approximately five times larger than for Kilauea. This explanation for the variations at Kilauea would require that the magmatic transfer is essentially one way: from Mauna Loa to Kilauea.

An alternative interpretation for the Kilauea helium isotopic variation is that the fluctuations represent changes in the source chemistry over this timescale. For example, in the case of continuous magma supply, the McKenzie & O'Nions (1991) percolation model might be applicable to individual magma batches. If this is the correct explanation for the variations shown in figure 5, then a new magma batch is produced every 100–300 years in the mantle beneath Kilauea. Based on the magma supply rate to Kilauea (roughly 7×10^6 m^3 month^{-1}; Dzurisin *et al.* 1984), this would require individual magma batches of approximately 1 km^3, which is roughly the predicted size of magma pulses (solitons) in the mantle (McKenzie & O'Nions 1991).

At present it is difficult to choose between the two hypotheses for the Kilauea helium isotopic variability. The relative sizes of the two volcanoes, coupled with the trace element evidence for magmatic exchange between them (Rhodes *et al.* 1989), suggests that magma mixing plays at least some role in the helium isotopic variability. Additional data will be required to confirm this. However, the data demonstrate that detailed studies of individual volcanoes may, in the future, place firm constraints on models for melting in mantle plumes.

In summary, the data presented in figures 4 and 5 demonstrate that isotopic variations within Hawaiian volcanoes can occur on extremely short timescales. With respect to the global isotopic compilations, the critical question is whether this type of variation is found at all ocean island volcanoes. Although there are too few isotopic studies of this kind, with the appropriate geological controls, to properly answer this question, existing data from other oceanic islands suggests that Hawaiian volcanoes are unique. Helium isotopic studies of Piton de la Fournaise (Reunion) demonstrated that the ^{3}He/^{4}He ratios (*ca.* $13 \times R_a$) in the erupted basalts have remained essentially constant for the past million years (Staudacher *et al.* 1990; Graham *et al.* 1990). Studies of the radiocarbon dated lava flows on Sao Miguel (Kurz *et al.* 1992), have demonstrated large variations over similar length scales as observed between Hawaiian volcanoes (i.e. *ca.* 20 km). However, in this case there is an isotopic gradient from one side of the island to the other, with little temporal control on the geochemistry of the basalts. There are many parameters which may differ between Reunion, the Azores, and Hawaii, including mantle composition, mantle structure, lithospheric thickness, depth of melting, and perhaps the extent of melting. The best constrained of these parameters is lithospheric thickness. Hawaii rests on relatively old oceanic crust and has thick lithosphere, which is circumstantial evidence that the lithosphere plays a key role in the temporal isotopic variations within Hawaiian volcanoes, as first suggested by Tatsumoto (1978).

5. Summary

New olivine/melt helium partitioning data presented here suggest that helium is an incompatible element with respect to melting, which contrasts with the conclusions of previous studies (e.g. Hiyagon & Ozima 1986). This supports the hypothesis that high ^{3}He/^{4}He ratios indicate undegassed mantle sources.

Data from the three Hawaiian volcanoes demonstrates how eruption age combined with isotope geochemistry can be used to place constraints on the melting dynamics beneath hotspots. A proper understanding of these local isotopic variations remains an important aspect of interpreting global isotopic variations, particularly in the context of mantle convection models. Helium isotopic data indicate little contribution of mantle plume material to present day eruptions of Hualalai and Mauna Loa volcanoes. In the past, both volcanoes have displayed higher ^{3}He/^{4}He ratios, which suggests that they have recently been removed from the actively melting hotspot mantle beneath Hawaii. In contrast, Kilauea has markedly higher ^{3}He/^{4}He ratios over the entire time sequence of subaerial eruptions, and dredged basalts from Loihi seamount have the highest ^{3}He/^{4}He ratios of any oceanic rocks. These data suggest a significant contribution from undegassed, plume-type mantle to Loihi and Kilauea, and restrict the size of the actively melting plume to a radius less than 40 km (roughly the distance between Loihi seamount and Mauna Loa), and greater than 30 km (the distance between Loihi seamount and Kilauea) (see figure 5). The complex temporal evolution of these volcanoes, illustrated in figure 4, is attributed here to heterogeneity within the mantle plume.

I acknowledge the collaboration of F. Frey, M. Garcia, D. Kammer, T. Kenna, R. Moore, C. Neal, M. Rhodes, N. Shimizu and F. Truesdell, and an extremely helpful review by F. Frey. Some of the work described here was supported by NSF grant OCE 90-817833. This is WHOI contribution number 8074.

References

Allegre, C. J., Staudacher, T., Sarda, P. & Kurz, M. 1983 Constraints on evolution of Earth's mantle from rare gas systematics. *Nature, Lond.* **303**, 762–766.

Allegre, C. J. & Turcotte, D. L. 1985 Geodynamic mixing in the mesosphere boundary layer and the origin of oceanic islands. *Geophys. Res. Lett.* **12**, 207–210.

Allegre, C. J., Staudacher, T. & Sarda, P. 1987 Rare gas systematics: formation of the atmosphere, evolution and structure of the Earth's mantle. *Earth planet. Sci. Lett.* **81**, 127–150.

Broadhurst, C. L., Drake, M. J., Hagee, B. E. & Bernatowicz, T. J. 1992 Solubility and partitioning of Ne, Ar, Kr, and Xe in minerals and synthetic basaltic melts. *Geochim. cosmochim. Acta* **56**, 709–723.

Bryan, W. B. & Moore, J. 1977 Compositional variations of young basalts in the Mid-Atlantic Ridge rift valley near 37 degrees N. *GSA Bull.* **88**, 556–570.

Chen, C. Y. & Frey, F. A. 1983 Origin of Hawaiian tholeiite and alkali basalt. *Nature, Lond.* **302**, 785–789.

Condomines, M., Gronvold, K., Hooker, P. J., Muehlenbachs, K., O'Nions, R. K., Oskarsson, J. & Oxburgh, E. R. 1983 Helium, oxygen, strontium and neodymium relationships in Icelandic volcanics. *Earth planet. Sci. Lett.* **66**, 125–136.

Craig, H. & Lupton, J. E. 1976 Primordial neon, helium and hydrogen in oceanic basalts. *Earth planet. Sci. Lett.* **31**, 369–385.

Davies, G. F. 1984 Geophysical and isotopic constraints on mantle convection: an interim synthesis. *J. geophys. Res.* **89**, 6017–6040.

Davies, G. F. 1990 Mantle plumes, mantle stirring and hotspot chemistry. *Earth planet. Sci. Lett.* **99**, 94–109.

Dzurisin, D., Koyanagi, R. Y. & English, T. T. 1984 Magma supply and storage at Kilauea Volcano, Hawaii. *J. Volcan. geotherm. Res.* **21**, 177–206.

Farley, K. A., Natland, J. H. & Craig, H. 1992 Binary mixing of enriched and undegassed (primitive?) mantle components (He, Sr, Nd, Pb) in Samoan lavas. *Earth planet. Sci. Lett.* **111**, 183–199.

Garcia, M. O., Kurz, M. D. & Muenow, D. 1990 Mahukona: the missing Hawaiian volcano. *Geology* **18**, 1111–1114.

Graham, D. W., Zindler, A., Kurz, M. D., Jenkins, W. J., Batiza, R. & Staudigel, H. 1988 He, Pb, Sr and Nd isotope constraints on magma genesis and mantle heterogeneity beneath young Pacific seamounts. *Contrib. Mineral. Petrol.* **99**, 446–463.

Graham, D., Lupton, J., Albarede, F. & Condomines, M. 1990 Extreme temporal homogeneity of helium isotopes at Piton de la Fournaise, Reunion Island. *Nature, Lond.* **347**, 545–548.

Griffiths, R. W. & Campbell, I. H. 1990 Stirring and structure in mantle starting plumes. *Earth planet. Sci. Lett.* **99**, 66–78.

Griffiths, R. W. & Campbell, I. H. 1991 On the dynamics of long-lived plume conduits in the convecting mantle. *Earth planet. Sci. Lett.* **103**, 214–227.

Gurnis, M. & Davies, G. F. 1986 The effect of depth-dependent viscosity on convective mixing in the mantle and the possible survival of primitive mantle. *Geophys. Res. Lett.* **13**, 541–544.

Hart, S. R. 1988 Heterogeneous mantle domains: signatures, genesis and mixing chronologies. *Earth planet. Sci. Lett.* **90**, 273–296.

Hart, S. R., Hauri, E. H., Oschmann, L. A. & Whitehead, J. A. 1992 Mantle plumes and entrainment: isotopic evidence. *Science, Wash.* **256**, 517–520.

Hiyagon, H. & Ozima, M. 1986 Partition of noble gases between olivine and basalt melt. *Geochim. cosmochim. Acta* **50**, 2045–2057.

Hofmann, A. W. & Hart, S. R. 1978 An assessment of local and regional isotopic equilibration in the mantle. *Earth planet. Sci. Lett.* **38**, 44–62.

Jenkins, W. J., Edmond, J. M. & Corliss, J. 1978 Excess ^{3}He and ^{4}He in Galapagos submarine hydrothermal waters. *Nature, Lond.* **272**, 156–158.

Kaneoka, I. 1983 Noble gas constraints on the layered structure of the mantle. *Nature, Lond.* **302**, 698–700.

Kellogg, L. H. & Wasserburg, G. J. 1990 The role of plumes in mantle helium fluxes. *Earth planet. Sci. Lett.* **99**, 276–289.

Klein, F. W. 1982 Patterns of historical eruptions at Hawaiian volcanoes. *J. Volcan. geotherm. Res.* **12**, 1–35.

Kurz, M. D. & Jenkins, W. J. 1981 The distribution of helium in oceanic basalt glasses. *Earth planet. Sci. Lett.* **53**, 41–54.

Kurz, M. D., Jenkins, W. J. & Hart, S. R. 1982a Helium isotopic systematics of oceanic islands: implications for mantle heterogeneity. *Nature, Lond.* **297**, 43–47.

Kurz, M. D., Jenkins, W. J., Schilling, J. G. & Hart, S. R. 1982b Helium isotopic variations in the mantle beneath the North Atlantic Ocean. *Earth planet. Sci. Lett.* **58**, 1–14.

Kurz, M. D., Jenkins, W. J., Hart, S. & Clague, D. 1983 Helium isotopic variations in Loihi Seamount and the island of Hawaii. *Earth planet. Sci. Lett.* **66**, 388–406.

Kurz, M. D., Garcia, M. O., Frey, F. A. & O'Brien, P. A. 1987 Temporal helium isotopic variations within Hawaiian volcanoes: basalts from Mauna Loa and Haleakala. *Geochim. cosmochim. Acta* **51**, 2905–2914.

Kurz, M. D., Colodner, D., Trull, T. W., Moore, R. B. & O'Brien, K. 1990 Cosmic ray exposure dating with *in situ* produced cosmogenic ^{3}He: results from young Hawaiian lava flows. *Earth planet Sci. Lett.* **97**, 177–189.

Kurz, M. D. & Kammer, D. P. 1991 Isotopic evolution of Mauna Loa Volcano. *Earth planet. Sci. Lett.* **103**, 257–269.

Kurz, M. D., Moore, R. B., Kammer, D. P. & Gulesserian, A. 1992 An isotopic study of dated alkali basalts from Sao Miguel, Azores: Implications for the Origin of the Azores Hot Spot. *Earth planet. Sci. Lett.* (Submitted.)

La Tourrette, T. Z. & Burnett, D. S. 1992 Experimental determination of U and Th partitioning between clinopyroxene and natural and synthetic basaltic liquid. *Earth planet. Sci. Lett.* **110**, 227–244.

Liu, M. & Chase, C. G. 1991 Evolution of Hawaiian basalts: a hotspot melting model. *Earth planet. Sci. Lett.* **104**, 151–165.

McKenzie, D. 1985 The extraction of magma from the crust and mantle. *Earth planet. Sci. Lett.* **74**, 81–91.

McKenzie, D. & O'Nions, R. K. 1991 Partial melt distributions from inversion of rare earth element concentrations. *J. Petrol.* **32**, 1021–1091.

Navon, O. & Stolper, E. 1986 Geochemical consequences of melt percolation: the upper mantle as a chromatographic column. *J. Geol.* **95**, 285–307.

O'Nions, R. K. & Oxburgh, E. R. 1983 Heat and helium in the Earth. *Nature, Lond.* **306**, 429.

Poreda, R., Schilling, J. G. & Craig, H. 1986 Helium and hydrogen isotopes in ocean-ridge basalts north and south of Iceland. *Earth planet. Sci. Lett.* **78**, 1–17.

Rhodes, J. M., Wenz, K. P., Neal, C. A., Sparks, J. W. & Lockwood, J. P. 1989 Geochemical evidence for invasion of Kilauea's plumbing system by Mauna Loa magma. *Nature, Lond.* **337**, 257.

Ribe, N. M. 1985 The generation and composition of partial melts in the Earth's mantle. *Earth planet. Sci. Lett.* **73**, 361–376.

Richter, F. M. 1986 Simple models for trace element fractionation during melt segregation. *Earth planet. Sci. Lett.* **77**, 333–344.

Riley, G. N. Jr & Kohlstedt, D. L. 1991 Kinetics of melt migration in upper mantle-type rocks. *Earth planet. Sci. Lett.* **105**, 500–521.

Rubin, M., Gargulinski, L. K. & McGeehin, J. P. 1987 Hawaiian radiocarbon dates. U.S. Geological Survey Prof. Paper 1350, 213–242.

Sarda, P. & Graham, D. 1990 Mid-ocean ridge popping rocks: implications for degassing at ridge crests. *Earth planet. Sci. Lett.* **97**, 268–289.

Silver, P. G., Calson, R. W. & Olson, P. 1988 Deep slabs, geochemical heterogeneity, and the large-scale structure of mantle convection: investigation of an enduring paradox. *A. Rev. Earth planet. Sci.* **16**, 477–541.

Sleep, N. H. 1990 Hotspots and mantle plumes: some phenomenology. *J. geophys. Res.* B **95**, 6715–6736.

Staudacher, T., Sarda, P. & Allegre, C. J. 1990 Noble gas systematics of Reunion island, Indian Ocean. *Chem. Geol.* **89**, 1–17.

Staudigel, H., Zindler, A., Hart, S. R., Leslie, T., Chen, C. Y. & Clague, D. 1984 The isotope systematics of a juvenile intraplate volcano: Pb, Nd, and Sr isotope ratios of basalts from Loihi seamount. *Earth planet. Sci. Lett.* **69**, 13–29.

Tatsumoto, M. 1978 Isotopic composition of lead in oceanic basalt and its implication to mantle evolution. *Earth planet. Sci. Lett.* **38**, 63–87.

Vance, D., Stone, J. O. H. & O'Nions, R. K. 1989 He, Sr, and Nd isotopes in xenoliths from Hawaii and other oceanic islands. *Earth planet. Sci. Lett.* **96**, 147–160.

White, W. M. 1985 Sources of oceanic basalts: radiogenic isotopic evidence. *Geology* **13**, 115–118.

Zindler, A. & Hart, S. R. 1986*a* Chemical geodynamics. *A. Rev. Earth planet. Sci.* **14**, 493–571.

Zindler, A. & Hart, S. R. 1986*b* Helium: problematic primordial signals. *Earth planet. Sci. Lett.* **79**, 1–8.

Melting study of a peridotite KLB-1 to 6.5 GPa, and the origin of basaltic magmas

By Eiichi Takahashi[1], Takeshi Shimazaki[1], Yasunori Tsuzaki[1] and Hideto Yoshida[2]

[1] *Earth and Planetary Sciences, Tokyo Institute of Technology, 2-12-1 Ookayama, Meguro-ku, Tokyo 152, Japan*
[2] *Geological Institute, Faculty of Science, Tokyo University, 7-2-1 Hongo, Bunkyo-ku, Tokyo 113, Japan*

With a newly established multi-anvil press in the Tokyo Institute of Technology, we have carried out a series of melting experiments on peridotite KLB-1 up to 6.5 GPa. Melt fractions of the peridotite were determined in a wide P–T range using extensive X-ray mapping analysis of run products by EPMA and a digitalized back-scattered electron image technique. Compositions of partial melts and solid residues were determined in the whole melting range up to 6.5 GPa. Given quantitative information on mantle melting, we discuss conditions of melting of various basalt magmas and the nature of their source materials. Our conclusions are consistent with the hypothesis that typical mid oceanic ridge basalts represent low pressure (*ca.* 1 GPa), low temperature ($T_p \approx 1300$ °C) partial melting products of mantle peridotite. Island arc picritic tholeiites may also be regarded as partial melts of a peridotitic source, at 1–2 GPa pressures and T_p ranging from 1400 to 1500 °C. However, proposed primary magmas for Hawaiian tholeiite are difficult to produce by partial melting of typical mantle peridotite at any depth under anhydrous conditions. Source materials for magmas in large hotspots (e.g. Hawaii, Iceland and some continental flood basalts (CFBS)) may be anomalously enriched in FeO and TiO_2 relative to typical upper mantle peridotites such as KLB-1.

1. Introduction

Experimental melting studies on mantle materials have been attempted by many workers to understand the origin of magmas in the Earth. Progress in recent numerical calculations concerning the dynamics of mantle materials and the physics of magma generation (McKenzie 1984; McKenzie & Bickle 1988; Watson & McKenzie 1991) has demonstrated the usefulness and need for such experimental data. We have carried out a series of melting experiments on mantle peridotites (Takahashi & Kushiro 1983; Takahashi & Scarfe 1985; Takahashi 1986a; Ito & Takahashi 1987). However, the phase diagrams currently available prove to be still incomplete, if numerical calculations of mantle melting dynamics are to be carried out. For example, there is no reliable information on the temperature dependence of partial melting at $P \geqslant 3$ GPa. Moreover, low pressure experimental data on melt fractions already published (Mysen & Kushiro 1977; Jaques & Green 1980) show significant discrepancies. Lack of detailed information on melt compositions at pressures above 3 GPa, sets limitations on the discussions of the origin of high MgO magmas.

Phil. Trans. R. Soc. Lond. A (1993) **342**, 105–120
Printed in Great Britain

The main purpose of this study is to determine an accurate phase diagram for mantle peridotite KLB-1 over a wide range of pressures and temperatures. Special care was taken to determine precisely pressures and temperatures for experiments above 4 GPa. Because accuracy in experimental temperatures is critically important, we first determined pressure corrections on thermocouple e.m.f.s. Pressure and temperature were then calibrated by measuring melting curves of gold and lead by a differential thermal analysis (DTA) method. To determine melt fractions in high pressure run products ($P \geqslant 3$ GPa), where partial melts crystallized completely during quenching (Takahashi 1986*a*, fig. 10), extensive X-ray mapping analysis was carried out. The present paper summarizes the above works (details will be published separately): (1) pressure corrections on thermocouple e.m.f.s (Tsuzaki & Takahashi 1993); (2) quantitative determination of melt fractions using X-ray mapping analysis (Yoshida & Takahashi 1993); and (3) phase relations of peridotite KLB-1 up to 15 GPa (Shimazaki & Takahashi 1993).

2. Experimental procedures

Methods and procedures for experiments below 3 GPa are similar to those reported in previous studies (Takahashi & Kushiro 1983). To reduce iron loss from the specimen and minimize chemical contamination from the container, a ribbon of Re foil was used to support the specimen in 1 atm† experiments. A Re foil was also used as an inner capsule instead of graphite in piston-cylinder experiments at 0.8 and 1.5 GPa. Compared with graphite, the redox state of the experimental charges inside the Re–Pt double capsules may be more oxidizing.

A multi-anvil press (SPI-1000) was designed by Takahashi and constructed by the Riken Co. in 1991 and installed in the ultrahigh-pressure experimental petrology laboratory of the Tokyo Institute of Technology. The SPI-1000 consists of a main frame, 1000 t uniaxial hydraulic ram and of a guideblock system to drive the multi-anvils. The multi-anvils are double staged and with newly designed bucket-type guideblocks to drive the cubic–octahedral anvils. Like double-stage multi-anvil presses in other laboratories (Ito *et al.* 1984), pressure media of various octahedral sizes can be used in SPI-1000 by changing the final truncation size of the tungsten carbide anvils. In the present experiments, an MgO pressure medium of 18 mm edge length (18 M) octahedron was used for the experiments at 4.6 and 6.5 GPa. The 18 M octahedron was compressed with tungsten carbide anvils using an edge length truncation of 11 mm. This assembly is suitable up to 11 GPa (see figure 4 for pressure calibration). Higher pressures can be achieved by reducing the pressure medium and truncation.

Both the heating and the hydraulic system of the SPI-1000 are operated by programmable controllers. Rates of compression and decompression, as well as heating and cooling, can be digitally programmed up to a pattern with 20 segments. In most experiments, total decompression time was programmed to be more than several hours (figure 1). In high pressure experiments, the peridotite specimens were kept at 1500 °C for 20–30 min, before reaching the desired melting temperature. This was done to let the subsolidus mineral assemblage adjust to the experimental pressure. Fluctuations in oil pressure and furnace temperature were of the order of ± 1 t or ± 3 °C. An example of a run record is given in figure 1.

† 1 atm $\approx 10^5$ Pa.

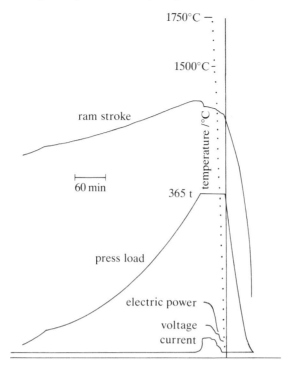

Figure 1. Run records for an experiment conducted at 6.5 GPa, 1750 °C and 20 min. In the SPI-1000 apparatus, both the hydraulic unit and the heating unit are controlled through a program and their fluctuations are very small (± 1 t and ± 3 °C).

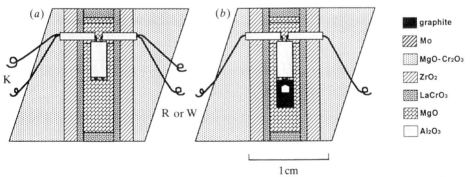

Figure 2. Furnace assemblies used for present experiments with the SPI-1000: (*a*) for thermocouple correction; (*b*) for melting experiment.

To avoid the temperature underestimation problems pointed out by Herzberg *et al.* (1990), the furnace assemblages shown in figure 2 were used in the present study (cf. Takahashi 1986*a*, fig. 2). To determine pressure effects on thermocouple e.m.f.s, the furnace assembly shown in figure 2*a* was used. Two sets of thermocouples were simultaneously used to record differences between their e.m.f.s at different pressures. K-, R- and W-type thermocouples (K, Chromel–Almel; R, Pt–Pt13% Rh; W, W5% Re–W26% Re) were tested. Apparent temperature differences between two thermocouple sets were recorded during heating and cooling cycles, at temperature intervals of 200 °C. The results obtained during the first heating cycles, at given

Phil. Trans. R. Soc. Lond. A (1993)

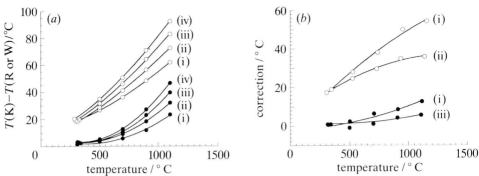

Figure 3. Experimental results for pressure corrections on thermocouple e.m.f.s. (*a*) Differences in apparent temperature readings with reference to the type-K thermocouple (Chromel–Almel). (i) 300 t; (ii) 400 t; (iii) 500 t; (iv) 600 t. (*b*) Absolute pressure corrections on e.m.f.s of the type-R (Pt Pt 13% Rh, —○—) and the type-W (W5% Re–W26% Re, —●—) thermocouples. (ɪ) 11.5 GPa; (ii) 5.7 GPa; (iii) 7.7 GPa.

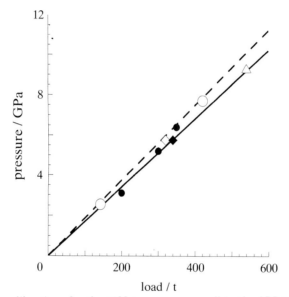

Figure 4. Pressure calibrations for the 18M type pressure cell in the SPI-1000 apparatus. Broken line is the room temperature calibration by Bi(I/II) and Bi(III/V) phase transitions. Solid line is the high temperature calibration determined with phase transitions of Fe_2SiO_4 (olivine/spinel) and SiO_2 (coesite/stishovite) at 1200 °C. Pressures determined by the melting curve of gold (●) by DTA (see figure 2*c*) coincide with the high temperature calibration line. ○, Bi; ◇, ◆, Fe_2SiO_4; △, SiO_2.

pressures, are plotted in figure 3*a*. Absolute pressure corrections for K-type thermocouple e.m.f.s., were calculated according to the method described by Getting & Kennedy (1970, and references therein). Absolute pressure corrections on R- and W-type thermocouple e.m.f.s (figure 3*b*) were calculated from the temperature differences (figure 3*a*) and the calculated absolute pressure corrections on K-type e.m.f.s. As e.m.f. pressure corrections for W-type thermocouples are very small, at least up to $P = 11.5$ GPa (figure 3*b*), in this study we used only W-type thermocouples.

Pressure calibrations were carried out with three independent methods (figure 4). Room temperature calibrations were established with phase transitions in Bi(I/II) and Bi(III/V) by resistance change. High temperature (1200 °C) calibrations were made by observing phase changes in Fe_2SiO_4 (olivine/spinel) and SiO_2 (coesite/stishovite) by quenching technique. High temperature calibration points are 9% lower in pressure generation efficiency than room temperature calibration.

To further check the accuracy of pressure and temperature calibration, we conducted a series of DTAS to determine melting curves for Au and Pb. Mirwald *et al.* (1975) determined melting curves for Au and Pb up to 6 GPa by a piston-cylinder apparatus. The pressure was calculated from the observed DTA signals obtained at given press loads. Pressure calibration data for Au are plotted on figure 4 and are located on the high temperature calibration curve established by phase transitions in Fe_2SiO_4 and SiO_2 at 1200 °C. Because melting curves for metals have small dT/dP slopes, temperature measurements must be very accurate to read pressures from the melting curve. The fit in figure 4, therefore, demonstrates that both pressure and temperature calibrations and their reproducibility in our multi-anvil system, are very accurate. The validity of our pressure corrections on thermocouple e.m.f.s is also supported. Levels of uncertainties in pressures and temperatures in the present experiments would be ± 0.1 GPa and ± 5 °C, respectively, well within typical uncertainty ranges in conventional piston-cylinder experiments.

3. The peridotite KLB-1

Starting material used in the present experiments is a powdered spinel lherzolite xenolith (KLB-1) already described by Takahashi (1986 *a*). Bulk chemical compositions of KLB-1 and well-documented mantle peridotite xenoliths (spinel lherzolite) are shown in figure 5. It is clear that KLB-1 represents a fertile example of typical mantle peridotites. Coherent variations in the major element chemistry of upper mantle peridotites (UMPS) have been recognized (Kuno & Aoki 1970; Maaløe & Aoki 1977) and interpreted as due to successive extraction of basalt magmas from the mantle. The least MgO-rich members of the UMPS (including KLB-1), therefore, may be closer to the primitive composition of the Earth's mantle (Ringwood 1975; Jagoutz *et al.* 1979; Hart & Zindler 1987).

Pyrolite compositions used in experimental melting studies (see following references) are also shown in figure 5. By definition, pyrolite compositions depend largely on the selected basalts for the mixing calculations (Ringwood 1975). A pyrolite composition based on Hawaiian tholeiite (Jaques & Green 1980) is very enriched in TiO_2. On the contrary, a MORB-pyrolite composition frequently used in recent experiments (Falloon & Green 1988; Falloon *et al.* 1988) is enriched in FeO compared with UMPS (figure 5).

In our study, partial melts from KLB-1 were obtained for a wide pressure and temperature range. Because KLB-1 is a fertile peridotite, whose bulk chemical composition falls in the well-defined compositional cluster of UMPS (figure 5), our experimental results may be applicable to the discussion of the petrogenesis of basaltic magmas under various tectonic settings, provided that they derive from typical UMPS and that their melting conditions are nearly anhydrous. In other words, the present experimental results may be useful to discuss diversities in source rock compositions of basaltic magmas.

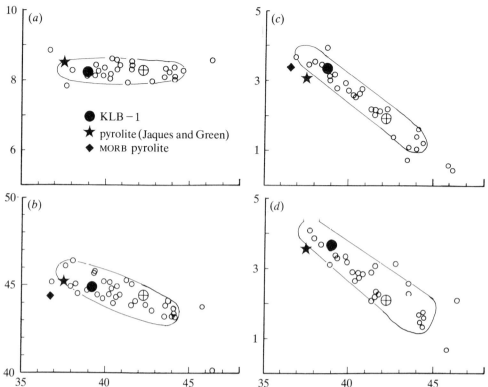

Figure 5. Bulk chemical compositions (in oxide wt%) of upper mantle peridotites (UMPs). See fig. 4 of Takahashi (1990) for data sources. Large circles with cross represent average composition of 384 spinel peridotite xenoliths by Maaløe & Aoki (1977). Composition of the starting material KLB-1 plots on the Fe-rich side of the well-defined compositional cluster of UMPs. In other words our experiments were conducted on a representative fertile composition of the Earth's upper mantle. Some pyrolite compositions experimentally tested are shown for comparison. (a) FeO; (b) SiO$_2$; (c) CaO; (d) Al$_2$O$_3$. Horizontal axes, MgO.

4. Melt fraction as a function of P and T

The degree of melting of mantle peridotite (ϕ) at given pressure and temperature $\phi\,(P, T)$ is one of the most important parameters to constrain the isentropic melting paths of uprising mantle materials (McKenzie 1984). At present, ϕ is determined only by experiments. Most experimental data on the melt fraction, however, are limited to low pressures, not greater than 1.5 GPa. Because of the lack of experimental data, McKenzie & Bickle (1988) assumed homologous relations on the $\phi(T)$ curve with pressure (see their fig. 6), i.e. they assumed that the degree of partial melting of a given peridotite does not change with pressure when temperature is normalized by the temperature intervals between the solidus and liquidus. This assumption is insufficient, because solidus melts change towards olivine rich compositions with increasing pressure (Takahashi & Kushiro 1983), i.e. the melting behaviour of mantle peridotite should become more eutectic with increasing pressure (Mysen & Kushiro 1977).

Melt fractions have been estimated by SEM (Jaques & Green 1980) and by β-track mapping analysis (Mysen & Kushiro 1977) of experimental run products. Mass

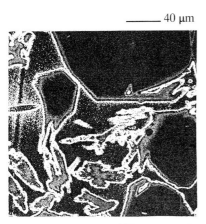

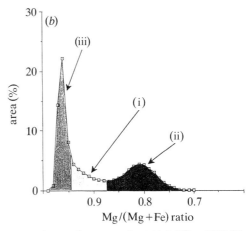

Figure 6. Example of X-ray mapping analysis on experimental run product (1.5 GPa, 1650 °C, 10.5 h). (*a*) Concentration map of Mg, consisting of large euhedral olivines (50–200 μm), small pyroxene quench crystals (10–50 μm) and glass. Olivines are mantled by thin rinds of overgrowth. Original photo was in 24 pseudo colours corresponding to MgO concentration. Approximately 30 % of the original area (400×400 μm^{-2}) is shown. (*b*) The Mg/(Mg + Fe) ratios of the 160 000 point analyses were computed by using a calibration line method. Quench crystals (i), overgrowth and glass (ii) can be distinguished from the stable olivine crystals (iii) by using the Mg/(Mg + Fe) ratios.

balance calculations may be useful especially at higher temperatures, where peridotites consist only of olivine and partial melt. Peridotite partial melts are unquenchable at pressures above about 2 GPa. Even at low pressures (when glass is formed in the run products) melt compositions are significantly modified due to overgrowth of adjacent solid phases (Jaques & Green 1979; Takahashi & Kushiro 1983; see also figure 6*a*).

To obtain reliable $\phi(P, T)$ data for KLB-1, we carried out an extensive digital X-ray mapping analysis of our run products, with a fully automated electron microprobe analyser JCMA-733MkII, at the University of Tokyo. Five spectrometers were used in fixed positions, to collect characteristic X-rays for Mg, Fe, Ca, Al and Cr. The X-ray countings were made for 40 ms for each point analysis and repeated at 1 μm grid intervals within a 400×400 μm^2 area, on the polished surface of our run products. To check spatial heterogeneities, the digital mapping was repeated at least at two locations within a single run product. With the aid of high speed stepwise motors for stage motion, the total counting time for the each 400×400 μm^2 area, was reduced to 3 h. To readily convert intensities into concentrations, we used a calibration line method for the above elements. Calibration lines were determined at the beginning of each analysis by carefully measuring 20 known standards mostly made out of peridotite constituent minerals. The resolution of our digital mapping analysis is good enough to identify very small changes in mineral composition (i.e. less than 0.3 mol % forsterite in olivine). The digital X-ray map is very useful in identifying stable mineral phases, overgrowth and quench crystals, glass and relics. An example of the MgO X-ray map of a run product obtained at 1.5 GPa and 1650 °C is shown in figure 6*a*. Modal proportions of olivine, quench crystals and glass, calculated from the Mg/(Mg + Fe) ratio for the same charge, are shown in figure 6*b*.

For experimental charges with uneven melt distributions, digital back-scattered electron image (BEI) analyses were carried out. In digital BEI, relative darkness of matter was converted into 16 pseudo colours, thereby used to distinguish quench

Table 1. *Run conditions, melt fractions and melt compositions*

P	T/°C	t/min	melt (%)	SiO₂	MgO	FeO	CaO	Al₂O₃	Na₂O	TiO₂
1 atm	1175	3330	5	—	—	—	—	—	—	—
1 atm	1204	2580	16	55.0	8.5	5.7	12.5	15.3	2.2	0.7
1 atm	1247	450	22	—	—	—	—	—	—	—
1 atm	1300	2100	31	54.9	15.0	7.8	8.8	12.0	0.6	0.4
1 atm	1350	360	37	—	—	—	—	—	—	—
1 atm	1400	360	45	51.2	22.5	9.1	6.5	9.5	0.4	0.3
1 atm	1500	60	49	49.6	24.4	10.3	6.5	7.9	0.2	0.3
1 atm	1600	420	64	47.8	29.5	9.4	5.4	6.5	0.1	0.2
0.8 GPa	1300	960	—	51.3	10.1	5.3	12.5	17.7	1.5	0.6
0.8 GPa	1350	240	—	50.9	11.3	6.5	11.6	17.0	1.4	0.6
0.8 GPa	1400	480	—	51.7	15.5	7.2	10.1	13.3	1.2	0.4
0.8 GPa	1450	60	—	49.8	16.6	7.4	9.8	14.1	1.2	0.4
0.8 GPa	1500	120	—	50.7	25.0	8.0	6.1	8.7	0.5	0.3
1.5 GPa	1400	140	17	—	—	—	—	—	—	—
1.5 GPa	1450	210	24	48.8	15.0	7.6	11.3	14.7	1.5	0.5
1.5 GPa	1500	40	43	—	—	—	—	—	—	—
1.5 GPa	1550	180	—	50.1	20.4	7.8	7.9	11.5	1.2	0.4
1.5 GPa	1600	150	50	—	—	—	—	—	—	—
1.5 GPa	1650	630	52	—	—	—	—	—	—	—
1.5 GPa	1700	36	56	—	—	—	—	—	—	—
3.0 GPa	1550	1020	—	47.0	19.2	7.8	12.2	11.0	1.2	0.9
4.6 GPa	1700	20	15	—	—	—	—	—	—	—
4.6 GPa	1750	20	22	47.6	19.8	11.3	7.8	10.2	1.1	0.6
4.6 GPa	1800	20	55	48.0	26.9	9.5	5.5	8.3	0.5	0.4
4.6 GPa	1850	10	65	47.2	34.3	8.4	3.7	5.2	0.3	0.2
4.6 GPa	1900	5	78	47.0	33.8	8.1	3.9	5.4	0.3	0.1
6.5 GPa	1850	20	23	—	—	—	—	—	—	—
6.5 GPa	1900	15	34	49.3	23.4	10.0	7.8	7.1	0.8	0.4
6.5 GPa	1950	15	72	49.0	29.4	8.7	4.9	6.8	0.1	0.2
6.5 GPa	2000	10	75	48.1	32.4	8.6	4.0	5.7	0.4	0.1

crystals from stable solid phases. Compared with the digital X-ray mapping, the resolution is poor (0.5 mol% in olivine solid solution) but a large area covering the entire run product (1 × 1 mm²) can be analysed in minutes.

Because the X-ray mapping and the digital BEI were taken on polished surfaces of run products, these analyses could be in error if melts and solids are unevenly distributed. This effect may become severe in experimental run products with a high degree of partial melting. We therefore conducted mass balance calculations using chemical analyses of melt and olivine and the bulk composition of KLB-1. The $\phi(P, T)$ curves for the peridotite KLB-1 were estimated from experimental results at 1 atm, 1.5 GPa, 4.6 GPa and 6.5 GPa (see table 1 and figure 7b). In the $\phi(T)$ curves obtained at 1 atm and 1.5 GPa, melt fraction changes significantly near both the solidus and the liquidus. Accordingly, there is a large temperature interval between $\phi = 0.4$ and 0.6 at low pressures. This melting behaviour is consistent with previous observations (McKenzie & Bickle 1988, fig. 6). The two-stage melting may be attributed to eutectic melting of pyroxenes and olivine near the solidus and disappearance of magnesian olivine near the liquidus. At pressures above about 3 GPa, however, this two-stage melting mode becomes less prominent (figure 7b).

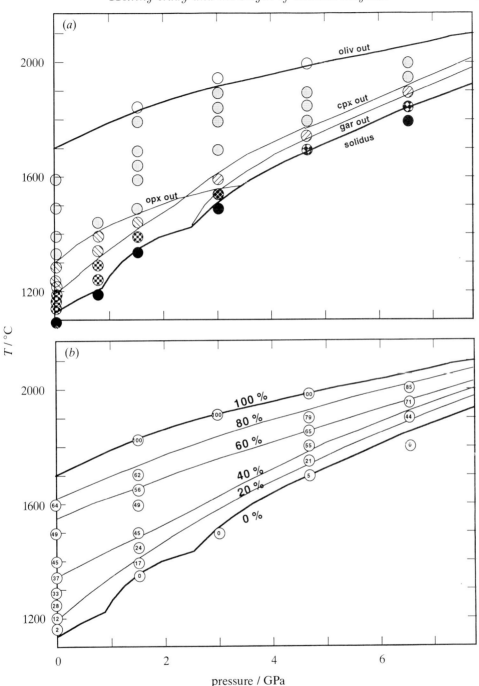

Figure 7. (a) Melting phase relations of peridotite KLB-1. All experimental data points at 1 atm, 0.8 GPa, 4.6 GPa and 6.5 GPa were newly obtained. Part of the data at 1.5 and 3 GPa were taken from Takahashi (1986a). (b) Melt fractions of peridotite KLB-1 determined by X-ray mapping analysis (see figure 6), digitalized REI techniques, and mass balance calculations. Eutectic behaviour become more prominent with increasing pressure (cf. figure 7b with figure 7a of McKenzie & Bickle (1988)).

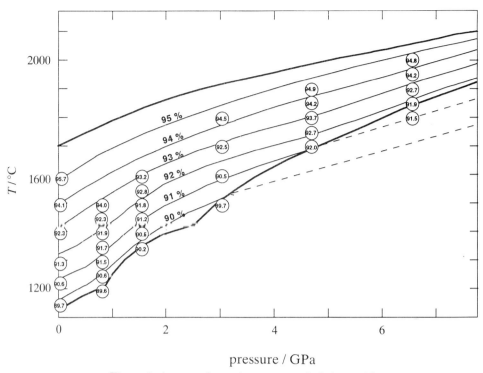

Figure 8. Average forsterite content of olivine residues.

5. Melt compositions as a function of P and T

Melting phase relations of the anhydrous peridotite KLB-1 are summarized in figure 7 a. Some experimental results at 1.5 GPa and 3.0 GPa taken from Takahashi (1986 a) are also reported. Run conditions and information on melt fractions and melt chemistry are summarized in table 1. Compositions of residual olivines are summarized in figure 7 b. Because of the presence of Fe-rich garnet in subsolidus conditions at $P \geqslant 3$ GPa, the Mg/(Mg + Fe) ratio of subsolidus olivine becomes higher than in the starting material (figure 8).

Partial melt compositions were determined by directly analysing quenched partial melts (either glass or quench crystals). For experiments at $P = 1$ atm, glass compositions obtained by EPMA (from large glass pods not less than 20 μm diameter) were cross-checked by using olivine–liquid Fe–Mg partitioning data ($K_D(\mathrm{Fe/Mg})^{\alpha\text{-liq}} = 0.30$ to 0.37 (Takahashi 1978)). Because of increasing overgrowth effects, glass compositions in the peridotite matrix obtained at 0.8 and 1.5 GPa are less reliable. Melt compositions obtained in large melt pods and satisfying the above K_D constraints are shown in table 2. To determine reliable melt compositions at 0.8 and 1.5 GPa, a series of experiments using the basalt/peridotite sandwich technique (Takahashi & Kushiro 1983) have been conducted. Additional basalt components (table 2) were carefully chosen not to alter the composition of peridotite matrix melts. A minimum quantity of basalt powder was added to the peridotite KLB-1 to obtain melt compositions similar to those of pure KLB-1. Compositions of melts resulting from the basalt/peridotite sandwich experiments, are compared with partial melt compositions of pure KLB-1 when direct EPMA analysis of matrix glass

Table 2. *Chemical compositions of peridotite partial melts*

(Compositions determined by direct EPMA analysis of quenched glass and those determined with the basalt/peridotite sandwich technique of Takahashi & Kushiro (1983).)

	0.8 GPa, 1325 °C, 18 h		0.8 GPa, 1500 °C, 2 h		starting basalts	
	KLB-1	KLB-1/FU12	KLB-1	KLB-1/T-16	FU12	T-16
SiO_2	51.4	51.2	50.4	48.7	46.8	46.5
TiO_2	0.5	0.5	0.3	0.3	0.6	0.5
Al_2O_3	15.0	14.6	8.6	8.1	15.0	14.2
FeO	6.9	7.0	7.9	8.3	9.1	7.4
MgO	12.6	13.5	24.9	26.7	13.1	15.9
CaO	11.9	11.3	6.1	6.2	14.0	13.3
Na_2O	1.2	1.3	0.5	0.5	1.5	1.6
K_2O	0.04	0.05	0.00	0.10	0.02	0.02

composition of olivine residues and Fe–Mg partition coefficients

Fo	91.5	91.2	94.2	93.9		
Kd(Fe/Mg)	0.30	0.33	0.34	0.37		

was successful (table 2). Agreements between melt compositions obtained by the two methods are very good in both major and minor elements (table 2).

In experimental run products obtained by the multi-anvil apparatus, partial melts were segregated within 10–20 min of run durations. This may be due to large melt fractions in most of run products (figure 7b). Larger temperature gradients (typically less than 50 °C mm^{-1}) in the multi-anvil cell (figure 2) compared with those in piston-cylinder cells (not more than 10 °C mm^{-1}) may be another factor in enhancing the melt segregation. The segregated melt compositions have been determined by broad-beam EPMA analysis.

Observed melt compositions are summarized in table 1 and some are plotted in figure 9. Because we have determined melt fractions $\phi(P, T)$ in figure 8b, and olivine is the only residual phase in the high temperature regions ($\phi \geqslant 0.4$, cf. figure 7a with b), melt compositions can be calculated by mass balance. Agreements between the calculated and measured results were acceptable except for Al_2O_3 and Na_2O. Disagreements may be due (at least in part) to uncertainty in the EDS X-ray analysis we used to determine the melt compositions. In fact, in EDS analysis, deconvolution of very small X-ray peaks for Al and Na from the dominant Mg peak may be difficult. Loss of Na during the EDS analysis was avoided by using relatively large beam sizes. In 1 atm runs, however, loss of Na during the experiments was significant and melt compositions are not reliable for sodium at $T \geqslant 1300$ °C.

6. Primary magma compositions and their origin

Given the $\phi(P, T)$ curves (figure 7b) and the melt compositions (figure 9) for typical upper mantle peridotite KLB-1, primary magma compositions and the magma production rate can be calculated for isentropic decompressional melting paths with various potential mantle temperatures (McKenzie 1984; McKenzie & Bickle 1988). If pressure effects on the $\phi(T)$ curves are taken into account, calculation schemes may become rather complicated. Before such numerical calculations are made, we discuss three representative melting régimes with reference to proposed primary magmas. In decompressional melting of mantle upwelling flow with $T_p = 1280$, 1550 and 1800 °C,

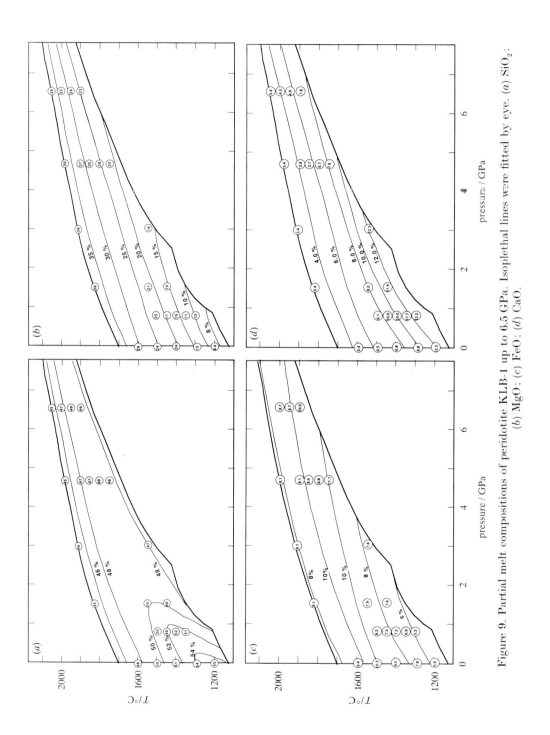

Figure 9. Partial melt compositions of peridotite KLB-1 up to 6.5 GPa. Isoplethal lines were fitted by eye. (a) SiO$_2$; (b) MgO; (c) FeO; (d) CaO.

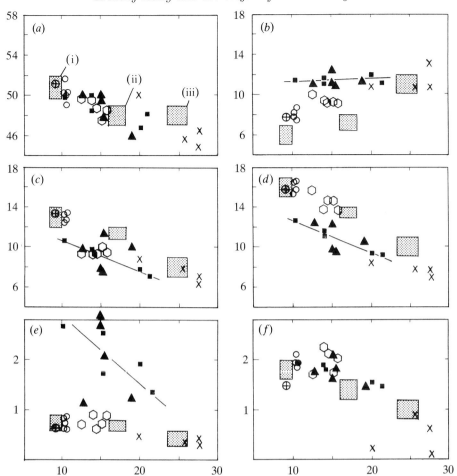

Figure 10. Expected initial melt compositions for decompressional melting of upper mantle peridotite for PMTs of (i) 1280, (ii) 1550 and (iii) 1800 °C are shown as shaded areas. Compositions of proposed primary magmas for MORBs (open circles), island arc tholeiites (open hexagons), Hawaiian tholeiites (solid squares), some continental flood basalts (solid triangles) and Archean komatiites (crosses) are shown. These rock compositions were taken from BVSP (1981), Takahashi (1986*b*) and Tatsumi *et al.* (1983). Large open circles with cross represent the average of 218 primitive MORB glass analyses by Melson *et al.* (1977). In oxide weight %, (*a*) SiO_2; (*b*) FeO; (*c*) CaO; (*d*) Al_2O_3; (*e*) TiO_2; (*f*) Na_2O. Horizontal axes, MgO.

partial melting of UMPs will take place at $\leqslant 1$ GPa, $\geqslant 4$ GPa, and $\geqslant 7$ GPa. Partial melting in the above régimes corresponds to the genesis of MORBs (McKenzie & Bickle 1988), Hawaiian tholeiite (Watson & McKenzie 1991) and Archean komatiites (Bickle *et al.* 1977; Takahashi 1990).

Expected initial melt compositions within the first $\geqslant 20$ % of batch partial melting under the above *P–T* conditions were estimated based on our new experimental results in table 1 and in figure 9. The estimated initial melt compositions are shown in figure 10. As proposed by McKenzie & Bickle (1988), primary MORB magma compositions plot very close to the expected melt compositions for $T_p = 1280$ °C. According to our experiments, the MgO content of the near solidus melt in this region is between 8 and 9 wt % (figure 9*b*), whereas estimated MORB primary magmas are slightly higher in MgO (10–11 wt %, see figure 10). On the other hand, the average of 218 primitive MORB glass analyses (open circle with cross in figure 10) by Melson *et al.* (1977) is identical to our expected composition except for minor differences in FeO and Na₂O. The deficiency in Na₂O in Melson *et al.*'s glass analyses may be due to Na volatilization during EPMA analyses. The higher FeO content in natural rocks with respect to experimental results may be explained by assuming that *ca.* 10% of iron in the MORB magmas is represented by ferric-iron but experiments were conducted under virtually ferric-iron free conditions due to the redox conditions imposed by the Re–Pt double capsule.

The genesis of MORB magmas is a fundamental problem in Earth sciences and, therefore, has been investigated by a great number of workers (e.g. Falloon & Green 1988; Kinzler & Grove 1992). We emphasize that the present study adds experimental support for the theory of MORB magma generation in relatively low pressure and temperature régimes (Green & Ringwood 1967; Presnall *et al.* 1979; Takahashi & Kushiro 1983; Fujii & Boughalt 1983; Presnall & Hoover 1984; Fujii & Scarfe 1985; Klein & Langmuir 1987; McKenzie & Bickle 1988).

Some primitive island-arc tholeiites are plotted in figure 10 (open hexagons). They plot in the compositional range of partial melts expected for T_p between 1280 and 1550 °C. Except for minor deficiencies in CaO and slight enrichment in Na₂O, their chemical signatures are very similar to those of MORB primary magmas (open circles). Estimated pressure and temperature conditions (1–2 GPa and *ca.* 1400 °C) for the primary island-arc tholeiite by Tatsumi *et al.* (1973) are consistent with our peridotite melting study. It follows that source materials for island-arc tholeiites may be peridotites similar to KLB-1.

Estimated primary basalt magma compositions for Hawaiian shield volcanoes differ among authors with different assumptions and different methods of calculation. Because primitive Hawaiian tholeiites are olivine saturated, proposed primary magmas for Kilauea and Mauna Loa (Irvine 1979; BVSP 1980; Wright 1984; Wilkinson 1985) lie essentially on an olivine fractionation line (figure 10*b–e*). Notable features for the Hawaiian tholeiites compared with MORBs are their higher FeO and TiO₂ contents. Even taking the highest possible MgO content for the Hawaiian primary magma (MgO $\approx$ 20 wt%), their TiO₂ contents are still two to three times higher than the expected composition of peridotite partial melts (figure 10*e*). Their FeO content might be compatible with our estimation if the highest MgO is assumed. However, their FeO is significantly higher than expected KLB-1 partial melts if the MgO of primary Hawaiian tholeiite is lower than 15 wt% (figure 10*b*). Furthermore, both CaO and Al₂O₃ for Hawaiian primary magmas are considerably lower than our estimated partial melts for all MgO concentrations (figure 10*c, d*). Accordingly, we propose that source materials for Hawaiian tholeiite magmas are significantly different from the normal UMP predominant in mantle xenoliths and tectonic blocks (figure 5). The anomalous nature of source rock compositions for Hawaiian tholeiite have been discussed by several authors (e.g. Wright 1984; Wilkinson 1985, 1991). Based on a quantitative olivine fractionation model (Takahashi 1986*b*), we have estimated that source materials for the Hawaiian mantle plume may be more than 50% enriched in iron compared with common UMPs (Takahashi & Uto 1993).

It is very important to note that proposed primary magmas for continental flood basalt terrains show chemical characteristics very similar to those of the Hawaiian hotspot (figure 10). This may indicate that mantle plume components of deep origin underneath very large hotspots may be different from shallower mantle peridotites (UMPs) in major element chemistry.

Stimulating and helpful discussions by Professor D. McKenzie were the motivation of this study. Critical readings of the manuscript by Professor D. McKenzie, Dr K. G. Cox and Dr G. Caprarelli at various stages of its preparation are greatly acknowledged. This is a contribution of the Department of Earth and Planetary Sciences, Tokyo Institute of Technology, no. 6. We acknowledge grants 02402019, 04216203 and 04201121 from the Ministry of Education, Science and Culture, Japan.

References

Bickle, M. J., Ford, C. E. & Nisbet, E. G. 1977 The petrogenesis of peridotitic komatiites: evidence from high pressure melting experiments. *Earth planet. Sci. Lett.* **37**, 97–106.

BVSP 1981 *Basaltic volcanism on the terrestrial planets.* (1286 pages.) New York: Pergamon.

Falloon, T. J. & Green, D. H. 1988 Anhydrous partial melting of peridotite from 8 to 35 kb and petrogenesis of MORB. *J. Petrol.: Oceanic and Continental Lithosphere* (ed. M. A. Menzies & K. G. Cox), pp. 379–414.

Falloon, T. J., Green, D. H., Hatton, C. J. & Harris, K. L. 1988 Anhydrous partial melting of a fertile and depleted peridotite from 2 to 30 kb and application to basalt petrogenesis. *J. Petrol.* **29**, 1257–1282.

Fujii, T. & Bougault, H. 1983 Melting relations of a magnesian abyssal tholeiite and the origin of MORBS. *Earth planet. Sci. Lett.* **62**, 283–295.

Fujii, T. & Scarfe, C. M. 1985 Composition of liquids coexisting with spinel lherzolite at 10 kb and the genesis of MORBS. *Earth planet. Sci. Lett.* **90**, 18–28.

Getting, I. C. & Kennedy, G. C. 1970 Effect of pressure on the emf of chromel-alumel and platinum–platinum 10% rhodium thermocouples. *J. appl. Phys.* **41**, 4552–4562.

Green, D. H. & Ringwood, A. E. 1967 The genesis of basaltic magmas. *Contrib. Mineral. Petrol.* **15**, 103–190.

Hart, S. R. & Zindler, A. 1986 In search of a bulk-earth composition. *Chemical Geol.* **57**, 247–267.

Herzberg, C. T., Gasparik, T. & Sawamoto, H. 1990 Origin of mantle peridotite: constraints from melting experiments to 16.5 GPa. *J. geophys. Res.* **95**, 15779–15803.

Irvine, T. N. 1979 Rocks whose composition is determined by crystal accumulation and sorting. In *The evolution of the igneous rocks* (ed. H. S. Yoder Jr), pp. 245–306. Princeton University Press.

Ito, E. & Takahashi, E. 1987 Melting of peridotite at uppermost lower mantle conditions. *Nature, Lond.* **328**, 514–517.

Ito, E., Takahashi, E. & Matsui, Y. 1984 The mineralogy and chemistry of the lower mantle: an implication of the ultrahigh-pressure phase relations in the system MgO–FeO–SiO_2. *Earth planet. Sci. Lett.* **67**, 238–248.

Jagoutz, E., Palme, H., Baddenhausen, H., Blum, K., Cendales, M., Dreibus, G., Spettel, B., Lorenz, V. & Wänke, H. 1979 The abundance of major, minor and trace elements in the earth's mantle as derived from primitive ultramafic nodules. *Proc. Lunar. Planet. Sci. Conf.* **10**, 2031–2050.

Jaques, A. L. & Green, D. H. 1979 Determination of liquid compositions in experimental, high pressure melting of peridotite. *Am. Mineral.* **64**, 1312–1321.

Jaques, A. L. & Green, D. H. 1980 Anhydrous melting of peridotite at 0–15 kb pressure and the genesis of tholeiitic basalts. *Contrib. Mineral. Petrol.* **73**, 287–310.

Kinzler, R. J. & Grove, T. L. 1992 Primary magmas of mid-ocean ridge basalts. Parts I, II. *J. geophys. Res.* **97**, 6885–6906; 6907–6926.

Klein, E. M. & Langmuir, C. H. 1987 Global correlation of ocean ridge basalt chemistry with axial depth and crustal thickness. *J. geophys. Res.* **92**, 8089–8115.

Kuno, H. & Aoki, K. 1970 Chemistry of ultramafic nodules and their bearing on the origin of basaltic magmas. *Phys. Earth planet. Inter.* **3**, 273–301.

Maaloe, S. & Aoki, K. 1977 The major element composition of the upper mantle estimated from the composition of lherzolites. *Contrib. Mineral. Petrol.* **63**, 161–173.

McKenzie, D. 1984 The generation and compaction of partially molten rock. *J. Petrol.* **25**, 713–765.

McKenzie, D. & Bickle, M. J. 1988 The volume and composition of melt generated by extension of the lithosphere. *J. Petrol.* **29**, 625–679.

Melson, W. G., Byerly, G. R., Nelson, J. A., O'Hearn, T., Wright, T. L. & Vallier, T. 1977 A catalogue of major element chemistry of abyssal volcanic glasses. *Smithson. Contrib. Earth Sci.* **19**, 31–61.

Mirwald, P. W., Getting, I. C. & Kennedy, G. C. 1975 Low-friction cell for piston-cylinder high-pressure apparatus. *J. geophys. Res.* **80**, 1519–1525.

Mysen, B. O. & Kushiro, I. 1977 Compositional variations of coexisting phases with degree of melting of peridotite in the upper mantle. *Am. Mineral.* **62**, 843–865.

Presnall, D. C., Dixon, J. R., O'Donnell, T. H. & Dixon, S. A. 1979 Generation of mid-ocean ridge tholeiite. *J. Petrol.* **20**, 3–35.

Presnall, D. C. & Hoover, J. P. 1984 Composition and depth of origin of primary mid-ocean ridge basalts. *Contrib. Mineral. Petrol.* **87**, 170–178.

Ringwood, A. E. 1975 *Composition and petrology of the Earth's mantle.* (618 pages.) New York: McGraw-Hill.

Takahashi, E. 1978 Partitioning of Ni^{2+}, Co^{2+}, Fe^{2+}, Mn^{2+}, and Mg^{2+} between olivine and silicate melts: compositional dependence of partition coefficient. *Geochim. cosmochim. Acta* **42**, 1829–1844.

Takahashi, E. 1986*a* Melting of a dry peridotite KLB-1 up to 14 GPa: implications on the origin of peridotitic upper mantle. *J. geophys. Res.* **91**, 9367–9382.

Takahashi, E. 1986*b* Origin of basaltic magmas: implications from peridotite melting experiments and an olivine fractionation model. *Bull. Volcanol. Soc. Japan* (30th Anniversary Issue), S17–S40. (In Japanese with English abstract.)

Takahashi, E. 1990 Speculations on the Archean mantle: missing link between komatiite- and depleted garnet peridotite. *J. geophys. Res.* **95**, 15941–15954.

Takahashi, E. & Kushiro, I. 1983 Melting of a dry peridotite at high pressures and basalt magma genesis. *Am. Mineral.* **68**, 859–879.

Takahashi, E. & Scarfe, C. M. 1985 Melting of peridotite to 14 GPa and the genesis of komatiite. *Nature, Lond.* **315**, 566–568.

Tatsumi, Y., Sakuyama, M., Fukuyama, H. & Kushiro, I. 1983 Generation of arc basalt magmas and thermal structure of the mantle wedge in subduction zones. *J. geophys. Res.* **88**, 5815–5825.

Watson, S. & McKenzie, D. 1991 Melt generation by plumes: a study of Hawaiian volcanism. *J. Petrol.* **32**, 501–537.

Wilkinson, J. F. G. 1985 Undepleted mantle composition beneath Hawaii. *Earth planet. Sci. Lett.* **75**, 129–138.

Wilkinson, J. F. G. 1991 Mauna Loan and Kilauean tholeiites with low ferromagnesian-fractionated 100 Mg/(Mg+Fe^{2+}) ratios: primary liquids from the upper mantle? *J. Petrol.* **32**, 863–907.

Wright, T. L. 1984 Origin of Hawaiian tholeiite: a metasomatic model. *J. geophys. Res.* **89**, 3233–3252.

Intershield geochemical differences among Hawaiian volcanoes: implications for source compositions, melting process and magma ascent paths

By F. A. Frey[1] and J. M. Rhodes[2]

[1] *Department of Earth, Atmospheric, and Planetary Sciences, Massachusetts Institute of Technology, Cambridge, Massachusetts 02139, U.S.A.*
[2] *Geology and Geography Department, University of Massachusetts, Amherst, Massachusetts 01003, U.S.A.*

As Hawaiian volcanoes develop, their lavas systematically change in composition and isotopic ratios of Sr, Nd and Pb. These trends provide important constraints for understanding plume-related volcanism as a volcano migrates away from the hotspot. There are also geochemical differences between Hawaiian shields. In particular, lavas from adjacent shields such as Kilauea and Mauna Loa on Hawaii and Koolau and Waianae on Oahu have significant differences in abundances of some major and incompatible elements and isotopic ratios of Sr, Nd and Pb. Some incompatible element abundance ratios, such as Zr/Nb and Sr/Nb, are correlated with intershield differences in Sr and Nd isotope ratios, but these isotopic ratios are not correlated with intershield differences in major element composition, or even parent/daughter abundance ratios such as Rb/Sr and Sm/Nd. Moreover, at Kilauea and Mauna Loa the intershield differences have apparently persisted for a relatively long time, perhaps 100 ka. These intershield geochemical differences provide important constraints on plume volcanism. Specifically, (i) each volcano must have distinct magma ascent paths from the region of melt segregation; (ii) the 25–50 km distance between adjacent, but geochemically distinct, shields requires that the sources vary on a similar scale, and that the melt production region is similarly restricted. The absence of correlations between lava compositions and radiogenic isotope ratios provides evidence for significant differences in melting process such as each shield forming by a different mean extent of melting with melt segregation at different mean pressures.

Two types of models are consistent with the intershield geochemical differences: (i) a relatively large radius, *ca.* 40 km, plume conduit with a systematic spatial distribution of geochemical heterogeneities; or (ii) a small radius, less than 20 km, plume conduit composed of geochemically distinct diapirs. Because relatively small radius diapirs of limited vertical extent are too small to create the large Hawaiian shields, a possible alternative is a continuous conduit containing solitary waves which transport geochemically distinct packets of material.

Phil. Trans. R. Soc. Lond. A (1993) **342**, 121–136

Printed in Great Britain

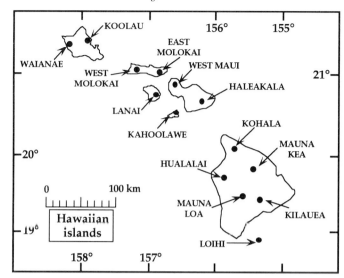

Figure 1. Location of the youngest Hawaiian volcanoes ranging from Waianae and Koolau on Oahu in the northwest to Loihi Seamount in the southeast (from Clague & Dalrymple 1987).

1. Introduction

The volcanoes comprising the Hawaiian Ridge formed as the Pacific plate migrated over a stationary magma source, a hotspot, that is created by a hot, upwelling mantle plume (Wilson 1963; Morgan 1971). Individual Hawaiian volcanoes evolve through well-characterized growth stages involving systematic temporal changes in lava composition (Stearns 1940; Clague & Dalrymple 1987; Frey & Roden 1987) and radiogenic isotope ratios of He, Sr, Nd and Pb (Chen & Frey 1985; Stille *et al.* 1986; West *et al.* 1987; Kurz & Kamner 1990). Most of the mass, 95–98 %, of Hawaiian volcanoes are formed during the shield-building stage as a consequence of relatively high eruption rates (Clague & Dalrymple 1987). Therefore, the shield-forming lavas are likely to provide the most direct information about the plume composition and dynamics. These shields are interpreted to be dominantly composed of tholeiitic basalt because such basalts form the surfaces of the active shields, Kilauea and Mauna Loa, and they are the major rock type in exposures of older, eroded Hawaiian shields.

　In this paper we focus on adjacent and approximately coeval Hawaiian shields composed of geochemically distinct tholeiitic basalts. These geochemical differences are manifested in major and trace element abundances as well as radiogenic isotope ratios. They provide information about the mantle source of Hawaiian shields and the magmatic processes associated with plume-related volcanism. Our objectives are (i) to evaluate the relative role of processes and differences in source in causing the intershield differences, and (ii) to identify the constraints on plume dynamics resulting from intershield geochemical differences.

　We discuss adjacent pairs of Hawaiian shields: Kilauea and Mauna Loa on the island of Hawaii and Koolau and Waianae on the island of Oahu (figure 1). Kilauea and Mauna Loa are actively growing shields, and there is an abundance of high quality geochemical data. The Koolau–Waianae pair is older, *ca.* 1.8–2.7 Ma for Koolau shield lavas and *ca.* 3.0–3.9 Ma for Waianae shield lavas (Langenheim & Clague 1987), and their formation may not have been coeval. This pair was chosen because Koolau lavas define an extreme in the range of major element compositions (Frey *et al.* 1993) and heavy radiogenic isotope ratios (Roden *et al.* 1984, 1993; West *et al.* 1987).

2. Intershield geochemical differences

(a) Major elements

At low pressures olivine is the liquidus phase of Hawaiian shield lavas, and the wide range in MgO content (*ca.* 5.5–23% in figure 2) reflects olivine accumulation and segregation (Powers 1955; Wright 1971). Intershield compositional differences are apparent in MgO variation plots (figure 2). We focus on lavas with more than 7% MgO because lavas with less than 7% MgO contain several phenocryst phases, and their compositions have been affected by post-melting segregation of clinopyroxene and plagioclase in addition to olivine; e.g. the abrupt changes of slope at *ca.* 6–7% MgO in the CaO–MgO and TiO_2–MgO trends (figure 2*a*). At more than 7% MgO, historic Kilauea and Mauna Loa lavas overlap in abundances of Al_2O_3 and Na_2O, but historic Kilauea lavas have higher CaO, TiO_2 and K_2O contents and lower SiO_2 contents (figure 2*a*). Because these intershield differences are not a function of MgO content, they represent differences in primary magma compositions (primary magmas have compositions that have not been changed by post-melting processes). Although abundances of major elements in older Hawaiian shield lavas are affected by low temperature alteration (Lipman *et al.* 1990; Frey *et al.* 1993), there are also important major element abundance differences between lavas from the adjacent Koolau and Waianae shields. Specifically, at *ca.* 7% MgO, Koolau lavas range to higher SiO_2 but lower TiO_2 and CaO contents (figure 2*b*). Relative to Kilauea and Mauna Loa lavas, Koolau lavas range to higher SiO_2 and Al_2O_3 and lower CaO contents (figure 2).

(b) Incompatible elements

In general, Kilauea lavas are more enriched in incompatible elements than Mauna Loa lavas (e.g. K_2O and TiO_2 in figure 2*a*; Rb, Sr, Zr, and Nb in figure 3*a*; La and Th, not plotted (also Tilling *et al.* 1987)). Especially good discriminants between historic Kilauea and Mauna Loa lavas are TiO_2 and Nb contents; in fact, TiO_2 is a better discriminant than the more incompatible oxides Na_2O, K_2O and P_2O_5 (figure 2*a*). Abundance ratios, such as K/Nb, Sr/Nb and Zr/Nb which are nearly uniform in genetically related basalts are also excellent intershield discriminants (figure 4). Utilizing such discriminants, Rhodes *et al.* (1989) found that some prehistoric lavas with Mauna Loa-like compositions were erupted on Kilauea from *ca.* 2.1 Ma to 300 years. Although a change in the Kilauea source composition could be inferred, Rhodes *et al.* (1989) favoured the interpretation that Kilauea's shallow magmatic plumbing system was occasionally invaded by magmas from Mauna Loa. An important observation is that the incompatible element abundance differences between historic Kilauea and Mauna Loa lavas are also characteristic of the oldest, subaerial lavas exposed at each volcano (e.g. Nb in figure 3*a*; Zr/Nb in figure 4). Although their ages are not well constrained, Hilina Basalt at Kilauea is *ca.* 31–100 ka (Easton *et al.* 1987) and Ninole Basalt at Mauna Loa is *ca.* 31–200 ka (Lipman *et al.* 1990). Therefore, these intershield differences have persisted for *ca.* 100000 years, a significant fraction of the shield growth stage.

Although at *ca.* 7% MgO most Waianae lavas have slightly higher TiO_2 and Nb contents than Koolau lavas, abundances of other incompatible elements such as Zr

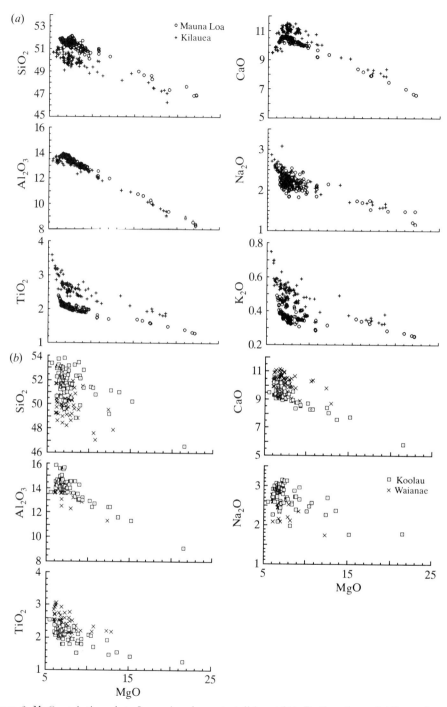

Figure 2. MgO variation plots for major elements (all in wt %). Rather than plotting only samples with a particular range in MgO content or correcting lava compositions to a common MgO content, we plot all of our data because (i) the effects of low pressure fractionation are clearly indicated at less than 7 % MgO, and (ii) it is evident that intershield differences are not a function of calculations that attempt to correct for the effects of post-melting, mineral-melt fractionation. (a) Comparison of historic Kilauea and Mauna Loa lavas (Rhodes, unpublished data). For lavas with more than

and Sr are similar in both shields (figures 2*b* and 3*b*); therefore, Zr/Nb and Sr/Nb abundance ratios in most Koolau lavas are higher than in most Waianae lavas (figure 4). Also relative to all Kilauea lavas and most Mauna Loa lavas, Koolau lavas have higher Zr/Nb and Sr/Nb (figure 4).

(c) *Isotopes*

Despite considerable isotopic variability within individual shields (West *et al.* 1987; Kurz & Kamner 1991; Roden *et al.* 1992), intershield differences in radiogenic isotopic ratios of Sr, Nd and Pb are large (figure 5). Moreover, isotopic data for the oldest lavas from Kilauea (Hilina Basalt) and Mauna Loa (Ninole Basalt) show that these isotopic differences have, like differences in incompatible element abundance ratios, persisted for *ca.* 100 ka. Note that Kilauea lavas have lower $^{87}Sr/^{86}Sr$ and higher $^{143}Nd/^{144}Nd$ than Mauna Loa lavas. Because Kilauea lavas have higher Rb/Sr (figure 4) and Nd/Sm (Budahn & Schmitt 1985; Tilling *et al.* 1987) than Mauna Loa lavas, these isotopic differences are inconsistent with the differences in parent/daughter abundance ratios.

The isotopic differences between Koolau and Waianae lavas are even larger. Koolau shield lavas are isotopically heterogeneous, but they define the high $^{87}Sr/^{86}Sr$, low $^{143}Nd/^{144}Nd$ and low $^{206}Pb/^{204}Pb$ extremes for Hawaiian lavas (Roden *et al.* 1993). In contrast, lavas forming the adjacent Waianae shield have $^{87}Sr/^{86}Sr$ and $^{143}Nd/^{144}Nd$ similar to Kilauea lavas but $^{206}Pb/^{204}Pb$ similar to Mauna Loa lavas (figure 5).

3. Interpretations

Although all studied Hawaiian shield lavas have Sr and Nd radiogenic isotope ratios between the fields for MORB and bulk Earth (West *et al.* 1987; Roden *et al.* 1993), the intershield isotopic differences are large (figure 5), and require that the sources of each shield contained isotopically distinct components, perhaps different proportions of the same components. Based on limited data for the oldest subaerial basalts from Kilauea and Mauna Loa (figure 5), these isotopic differences persist for a long time, perhaps at least 100 ka. Among Hawaiian shield lavas, those from Koolau volcano define an isotopic extreme (relatively low $^{143}Nd/^{144}Nd$ and $^{206}Pb/^{204}Pb$ and high $^{87}Sr/^{86}Sr$ (Roden *et al.* 1993)), and they are also extreme in several major element characteristics such as high SiO_2, low CaO (figure 2) and low total iron contents (not plotted). However, there is no systematic correlation between isotopic ratios of Sr, Nd and Pb and major element abundance characteristics. Specifically, West *et al.* (1987) found that lavas from Kahoolawe Volcano are isotopically similar to Koolau lavas, but Kahoolawe shield lavas do not

---

7% MgO there are distinct intervolcano differences for TiO_2, K_2O, CaO and SiO_2. (*b*) Comparison of Waianae and Koolau basalts. The most distinct differences are in SiO_2 and CaO. K_2O data are not shown because of K_2O loss caused by low-temperature alteration. (Koolau data from Frey *et al.* (1993); Waianae data for samples with more than 6% MgO from Macdonald & Katsura (1964), Zbinden & Sinton (1988), Sinton & Presley unpublished). All abundance data for lavas from Kilauea, Mauna Loa and Koolau volcanoes in this and subsequent figures were determined by a combination of X-ray fluorescence at the University of Massachusetts and instrumental neutron activation analysis at the Massachusetts Institute of Technology. Therefore, the compositional differences between these shields are not a result of systematic interlaboratory errors. Abundance data for Al_2O_3 and Na_2O in Waianae lavas reported by Macdonald & Katsura (1964) were not plotted because they are systematically different from more recent data for Waianae lavas.

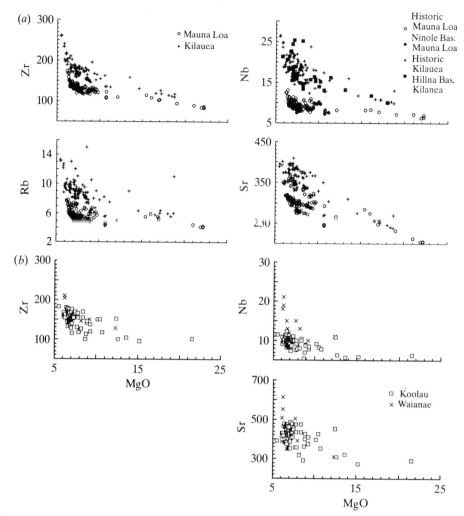

Figure 3. Abundances of the incompatible elements of Rb, Sr, Zr and Nb (all in p.p.m.) against MgO (wt %). (*a*) Comparison of historic Kilauea and Mauna Loa lavas (Rhodes, unpublished data). The Nb panel includes data for the oldest basalts from Kilauea (Hilina) and Mauna Loa (Ninole) (Ninole data from Lipman *et al.* (1990); Hilina data from Chen, Frey and Rhodes, unpublished). (*b*) Comparison of Waianae and Koolau basalts (Koolau data from Frey *et al.* (1993); Waianae data from Sinton & Presley, unpublished, and Feigenson, unpublished.) Rb data not shown because of the Rb loss caused by alteration of old shield lavas.

have the distinctive major element compositions that characterize Koolau lavas (Frey *et al.* 1993). In fact, in several respects, such as relatively high SiO_2 content, Koolau and Mauna Loa lavas are similar (greater than 51 % SiO_2 at 7 % MgO, figure 2), but they are isotopically very different (figure 5). Consequently, we conclude that the intershield differences in major element composition were not controlled by the components creating intershield isotopic differences.

Do the intershield differences in major element compositions reflect differences in process rather than differences in source composition? Klein & Langmuir (1987) used abundances of SiO_2, total iron, Al_2O_3 and Na_2O in regionally averaged mid-ocean ridge basalt (MORB) combined with experimental results for partial melting of spinel

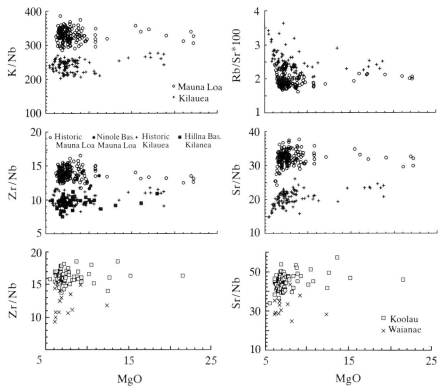

Figure 4. Incompatible element abundance ratios against MgO content (wt %). Upper four panels: historic lavas from Kilauea and Mauna Loa. The Zr/Nb panel includes data for the oldest basalts from Kilauea (Hilina) and Mauna Loa (Ninole). Lower two panels: Waianae and Koolau. Because K and Rb abundances are affected by post-magmatic alteration, K/Nb and Rb/Sr data are not shown. Data sources as indicated in caption to figure 3.

peridotite to argue for regional differences in the mean depth and extent of melt segregation. Melts segregating from a mantle plume should also reflect a range in pressures and extents of melting (Ribe 1988; Wyllie 1988; Liu & Chase 1991 a). However, based on both geophysical and geochemical arguments, Hawaiian tholeiitic magmas segregate at higher pressures than MORB. For example, Watson & McKenzie (1991) developed a numerical model of the Hawaiian plume that is consistent with the observed melt production rate, residual depth anomaly and geoid anomaly. They concluded that melt production occurs between depths of 82–136 km; consequently, Hawaiian shield lavas segregate at pressures which encompass the transition region from spinel- to garnet-peridotite (McKenzie & O'Nions 1991). Because garnet has high contents of Al_2O_3, Sc and heavy REE relative to coexisting melt, the lower Al_2O_3, Sc and heavy REE contents of Hawaiian shield lavas relative to MORB are consistent with garnet as an important residual phase during generation of Hawaiian shield lavas. The uniformity of heavy REE contents in Hawaiian tholeiites (when adjusted to a common MgO content) is also a strong argument for residual garnet (Hofmann *et al.* 1984).

As inferred by Klein & Langmuir (1987) and shown by Takahashi *et al.* (this symposium), within the stability field of spinel peridotite the SiO_2 contents of partial melts decrease with increasing pressure. Although the effect of pressure on the SiO_2

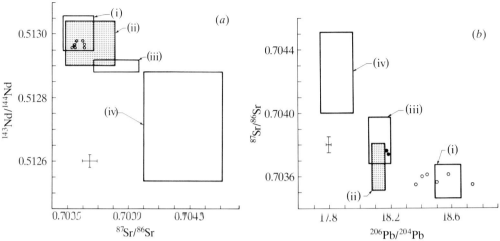

Figure 5. ^{143}Nd/^{144}Nd against ^{87}Sr/^{86}Sr and ^{87}Sr/^{86}Sr against ^{206}Pb/^{204}Pb showing the large differences between the adjacent shields. Mauna Loa–Kilauea and Waianae–Koolau. (a) (i) Historic Kilauea (24, 21), (ii) Waianae (16, 16), (iii) Mauna Loa (33, 5), (iv) Koolau (23, 17). (b) (i) Historic Kilauea (24, 5), (ii) Waianea (16, 4), (iii) Mauna Loa (33, 30), (iv) Koolau (23, 11). Numbers within parentheses indicate number of samples analysed with number for ^{87}Sr/^{86}Sr given first. Note that the oldest lavas from Mauna Loa, Ninole Basalt (two filled circles in right panel), lie within the field for younger Mauna Loa lavas. Also, the oldest lavas from Kilauea, Hilina Basalt (open circles in both panels), lie within or extend slightly the fields for younger Kilauea lavas. Another important point is that Mauna Loa lavas have higher ^{87}Sr/^{86}Sr than Kilauea lavas but they have lower Rb/Sr (figure 3a). Error bars in lower left indicate typical two sigma errors. Mauna Loa and Kilauea data from O'Nions et al. (1977), Tatsumoto (1978), White & Hofmann (1982), Hofmann et al. (1984), Stille et al. (1986), Kurz & Kamner (1991), Chen, Frey and Rhodes (unpublished data for Hilina basalt). Waianae data from White & Hofmann (1982), Stille et al. (1983), and Feigenson (unpublished). Koolau data from Stille et al. (1983) and Roden et al. (1993).

contents of melts in equilibrium with garnet peridotite is not as well established, the relatively higher abundances of SiO_2 in Mauna Loa and Koolau lavas (figure 2) may reflect melt segregation at a lower mean pressure than lavas from their respective adjacent volcanoes, Kilauea and Waianae. Total iron contents in partial melts increase with increasing pressure (Klein & Langmuir 1987; Takahashi et al., this symposium), therefore, the atypically low total iron contents of Koolau lavas (generally lower than 11.7 % Fe_2O_3, as total iron (Frey et al. 1993)) are also consistent with segregation at relatively low pressure.

The higher K_2O and TiO_2 contents of Kilauea lavas (figure 2a) are accompanied by higher abundances of incompatible trace elements (figure 3a). If the sources of Kilauea and Mauna Loa lavas were similar in incompatible element content, Kilauea lavas were derived by lower extents of melting (F). In this context, the similar Al_2O_3 and Na_2O contents of Kilauea and Mauna Loa lavas at a given MgO content (figure 2a) are surprising, because in regionally averaged MORB, these oxides have been used as sensitive indicators of F (Klein & Langmuir 1987). However at the higher pressures of melt segregation inferred for Hawaiian shield lavas, Al_2O_3 and probably Na_2O become more compatible in residual minerals, and their abundances are not as sensitive to extent of melting.

Abundances of CaO are also a good intershield discriminant; e.g. at a given MgO content, CaO decreases in the order Kilauea > Mauna Loa ≈ Waianae > Koolau (figure 2). Within the stability field of spinel peridotite, CaO contents of melts increase with F until clinopyroxene is exhausted from the residue (Kinzler & Grove 1992). If a similar trend is valid at higher pressures, the relatively high CaO content of Kilauea lavas is inconsistent with the inference based on incompatible element abundances that Kilauea lavas formed by a lower mean extent of melting than Mauna Loa lavas. This paradox is resolved, if the Kilauea source had relatively higher abundances of incompatible elements. It is also possible that the CaO contents of melts in equilibrium with clinopyroxene and garnet are sensitive to pressure in the 25–45 kb range; however, there are insufficient experimental data to evaluate this hypothesis. Another possibility is that CaO abundances reflect differences in volatile content (Wyllie 1988). In particular, the effect of CO_2 is to lower SiO_2 and increase CaO in partial melts (Eggler 1978); possibly, Kilauea lavas were derived from a source with higher CO_2 content (Garcia *et al.* 1989).

In summary, the limited experimental data for the compositions of partial melts in equilibrium with garnet peridotite preclude confident interpretations of intershield differences in major element composition. However, it is possible that these compositional differences reflect variations in mean depth and extent of melting, and perhaps variations in volatile content.

A difficulty in using lava composition to constrain melting processes is that erupted magmas are probably mixtures of melts formed over a range in pressure and extent of melting (Thompson 1987). Eggins (1992*a*) concluded that picritic Hawaiian tholeiites with *ca.* 16% MgO (Clague *et al.* 1991) are not in equilibrium with garnet peridotite at any pressure or temperature, but that they are in equilibrium with harzburgite at 20 kb. How can this constraint be consistent with control of some trace element abundances by residual garnet? Eggins (1992*b*) inferred that the rate of melting must decrease as the plume approaches the lithosphere. Therefore he proposed that the bulk of the shield lavas formed by relatively low F in the presence of garnet, but because of equilibration during ascent by porous flow the major element compositions reflect equilibration with harzburgite at the top of the melting column. Perhaps the incompatible element-rich Kilauea lavas contain more of these low F melts than Mauna Loa lavas. Because the extent of melting in a plume must decrease radially from the relatively high temperature plume axis toward the relatively cool periphery, a possible explanation is that the source of Kilauea lavas is offset from the plume axis. If the relatively low SiO_2 contents of Kilauea lavas reflect melt segregation at higher pressures, then magmas derived from the plume periphery reflect a lower F and higher pressure of melt segregation.

McKenzie & O'Nions (1991) proposed a different interpretation for the relative incompatible element enrichment of Kilauea lavas. They proposed that as a Hawaiian shield ages, partial melts of asthenosphere percolate through and interact with the overlying plume matrix. During this process abundance ratios among incompatible elements are changed because of chromatographic effects. This model was proposed to explain the lower abundances of light REE and other incompatible elements such as Sr (figure 3*a*) in Mauna Loa lavas relative to Kilauea lavas. However, this model predicts that Mauna Loa lavas should be more MORB-like in isotopic ratios; consequently, the model does not satisfactorily explain the higher $^{87}Sr/^{86}Sr$ and lower $^{143}Nd/^{144}Nd$ of Mauna Loa lavas (figure 5). In addition, West *et al.* (1987) concluded that the $^{87}Sr/^{86}Sr$–$^{206}Pb/^{204}Pb$ trend defined by Hawaiian shield lavas is inconsistent with an important role for a MORB-like component. Moreover, the McKenzie & O'Nions model implies that the youngest shield lavas on extinct Hawaiian shield should also have relatively low abundances of incompatible elements, but this is not observed (Leeman *et al.* 1980; Budahn & Schmitt 1985). Therefore, this model does not satisfactorily explain intershield geochemical differences.

Abundance ratios of incompatible elements are relatively insensitive to fractional crystallization, and it is unlikely that the intershield differences in K/Nb, Zr/Nb and Sr/Nb (figure 4) reflect shallow processes, such as fractionation and mixing in a replenished magma chamber (Loubet *et al.* 1988; Nielsen 1990). In particular, ratios such as Zr/Nb and Sr/Nb are correlated with $^{87}Sr/^{86}Sr$; e.g. all decrease in the order Koolau > Mauna Loa > Waianae $\geqslant$ Kilauea (figures 4 and 5); therefore, some of intershield variations in incompatible element abundances may reflect source differences. However, there is evidence that intershield differences in incompatible element abundance ratios do not reflect long-term differences in sources. For example, relative to Mauna Loa lavas, Kilauea lavas have higher Rb/Sr but lower $^{87}Sr/^{86}Sr$ and lower Sm/Nd but higher $^{143}Nd/^{144}Nd$ (figures 4 and 5). This decoupling between parent/daughter ratio and isotopic ratio cannot be a long-lived effect and requires a recent change in parent/daughter abundance ratios caused by processes such as partial melting, mixing and metasomatism. The absence of a correlation between parent/daughter abundance ratio and isotopic ratio is even more dramatic for Koolau volcano which has unusually low $^{143}Nd/^{144}Nd$ (figure 5) but Sm/Nd ratios that encompass the range of Kilauea and Mauna Loa lavas (Frey *et al.* 1993).

4. Implications

During the growth of a Hawaiian volcano there are important temporal trends in geochemical characteristics. The transition from the shield stage to post-shield and rejuvenated stages has been studied at several volcanoes, and the geochemical trends have provided important constraints for understanding the evolution of Hawaiian volcanoes as they move away from the hotspot. Although the geochemical characteristics of lavas forming a shield may also vary systematically as a function of eruption age (Rhodes 1983; Kurz & Kamner 1991), we emphasize that lavas from adjacent and approximately coeval Hawaiian volcanoes (Mauna Loa and Kilauea on Hawaii, Koolau and Waianae on Oahu) have significant differences in major and incompatible element abundances and radiogenic isotope ratios, and the differences between Mauna Loa and Kilauea have persisted for a significant fraction of shield stage growth. Moreover, most intershield geochemical differences are larger than known intrashield geochemical variations.

The systematic geochemical differences between the lavas forming adjacent shields lead to constraints on the processes involved in creation of Hawaiian shields, and they have the following important implications for plume models.

(*a*) Each volcano must have distinct magma ascent paths. This requirement has not been recognized from the locations of earthquakes at depths of more than 40 km (Klein 1987).

(*b*) The 25–50 km distance between adjacent calderas requires that the source regions of Hawaiian volcanism vary on a similar scale. Although geochemical heterogeneity on this length scale is not surprising, the large volumes of Hawaiian shields requires that substantial volumes of source material remain geochemically distinct, possibly throughout growth of the shield stage, *ca.* 0.5–1 Ma. Extraction of melts from large plumes, more than 50 km in diameter, cannot explain the intershield differences unless the plume has a systematic distribution of hetero-geneities. If such heterogeneities exist in the source, they may be preserved in the overlying volcanoes (Eggins 1992b) because melts ascend nearly vertically in the fluid dynamic model for plumes proposed by Ribe & Snooke (1987). Entrainment of asthenospheric wallrocks might also create heterogeneity on an intershield scale (Griffiths & Campbell 1991), but the isotopic and incompatible element charac-teristics of adjacent shields, such as Kilauea and Mauna Loa, do not indicate that one shield contains more of an asthenospheric, MORB-related component.

(c) Because intershield differences in radiogenic isotopic ratios are not correlated with abundances of major elements, there is also evidence for differences in the melting process. For example, intershield differences in SiO_2 and total iron contents (e.g. relatively high SiO_2 and low total iron in Koolau lavas), may reflect differences in mean pressure of melt segregation. In addition, the absence of correlations between Rb/Sr and $^{87}Sr/^{86}Sr$ and Sm/Nd and $^{143}Nd/^{144}Nd$ indicate that the parent/daughter abundance ratios were affected by the partial melting process (e.g. different mean extents of melting for each shield) or the source compositions were recently affected by an enrichment (metasomatic?) process. Also, the intershield differences in incompatible element abundances are not consistent with the order of incompatibility typically inferred for partial melts of spinel peridotite (Sun & McDonough 1989, fig. 1). For example, Ti is a better intershield discriminant than Na or P, and Nb is a better discriminant than K or Rb (figures 2 and 3). Either the sources have variable Ti and Nb contents or the transition from partial melting of spinel peridotite to garnet peridotite increases the relative compatibility of K, Rb, Na and P. Unfortunately, rigorous assessment of how much of the intershield compositional differences reflect variations in pressure of melt segregation is not yet possible, because there are insufficient experimental data for melts in equilibrium with garnet peridotite.

5. Structure of the Hawaiian plume

Possible explanations for intershield differences in isotopic ratios of Sr, Nd and Pb are: (i) shallow mixing processes involving plume and oceanic lithosphere components; (ii) geochemical heterogeneities within a large plume, either intrinsic or resulting from entrainment of wallrocks during ascent; or (iii) relatively small, closely spaced, geochemically distinct diapirs with each shield forming from a different diapir.

In a $^{87}Sr/^{86}Sr$–$^{206}Pb/^{204}Pb$ plot, data for Hawaiian shield lavas do not trend toward the MORB field; therefore, West *et al.* (1987) concluded that shield lavas do not contain a significant amount of oceanic lithosphere or entrained MORB-related asthenosphere. However, a plume conduit, *ca.* 40 km radius, with systematic spatial geochemical heterogeneities could explain the intershield differences, especially if the melt ascent paths are nearly vertical (Ribe & Smooke 1987). If these spatial heterogeneities are inherited from the boundary layer source, it is surprising that these small-scale variations can be preserved during ascent over 100s of kilometres. An alternative is that entrainment of wallrocks, perhaps lower mantle (Hart *et al.* 1992), creates the spatial geochemical heterogeneity. If interaction of the plume conduit with the lithosphere results in tilting of the conduit and asymmetry in the extent of melting contours, then intershield differences in mean extent and pressure of melting are expected (figure 6a, b).

Phil. Trans. R. Soc. Lond. A (1993)

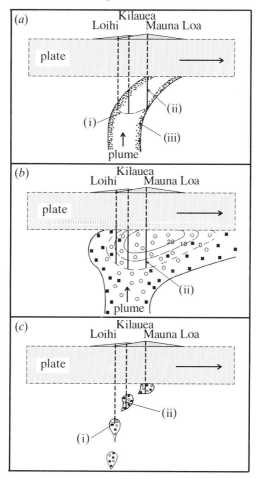

Figure 6. Plume models that could explain the observed intershield geochemical differences. In each sketch, the oceanic lithosphere (plate) is 70 km thick and melting initiates at a depth of 130 km (Watson & McKenzie 1991). (i) Beginning of melting; (ii) melting column; (iii) entrained mantle. In (a) the plume conduit has relatively cool margins which are a mixture of plume material and entrained wallrock. Within the conduit, melting initiates at the solidus labelled (i). Some of the intershield geochemical differences are a result of conduit tilting at shallow depths, i.e. within 70 km of the lithosphere–asthenosphere contact. Because the centre of the conduit is at a higher temperature than the margins, lavas derived from the centre (Kilauea shield) segregate at a higher mean pressure than lavas derived from the margin (Mauna Loa shield). However, the melting column for Mauna Loa lavas is longer than that for Kilauea lavas; therefore, Mauna Loa lavas reflect a larger, mean extent of melting, consistent with their lower abundances of incompatible elements relative to Kilauea lavas. An implication of this model is that the geochemical characteristics of a shield should change as it moves from the position of Loihi to Mauna Loa. In (b) the plume configuration is similar to (a) except that the plume conduit contains intrinsic geochemical heterogeneities (indicated by the varying proportion of open circles and closed squares) and interaction between the plume conduit and lithosphere is more extensive resulting in distortion of the axisymmetric extent of melting contours (labelled 0, 10 and 20 in figure) illustrated in fig. 12 of Watson & McKenzie (1991). Mauna Loa lavas reflect a larger mean extent of melting than Kilauea lavas. In (c) the plume is composed of discrete diapirs which may reflect discontinuous input from the plume source or transformation of a continuous conduit into discrete diapirs as the plume ascends and interacts with the mantle (Liu *et al.* 1991). An attractive aspect of this model is that geochemical differences between shields are expected because each diapir could be different in composition (indicated by variable proportion of open circles and closed squares),

An alternative hypothesis is that the plume conduit is discontinuous and is defined by a series of distinct diapirs. For example, Skilbeck & Whitehead (1978) proposed that tilting of the plume conduit may be sufficient to create diapiric instability. Although Griffiths & Richards (1989) used the sharpness of the bend in the Hawaiian Emperor chain to conclude that the tilt of the Hawaiian plume conduit is not sufficient to cause diapiric instability, Whitehead & Helfrich (1990) note that this conclusion is dependent on the assumed radius of the plume conduit and the viscosity of the surrounding mantle. Moreover, there are other mechanisms for creating a non-continuous plume conduit (Olson 1990). For example, the supply of buoyant material in the plume source may be episodic rather than continuous, or diapir chains can develop as a continuous conduit passes through a phase change during ascent (Liu *et al.* 1991). From a geochemical perspective, creation of each shield from an individual diapir is attractive because in addition to likely intrinsic geochemical differences between diapirs, differences in size and temperature of each diapir would lead to differences in mean extent and depth of melting. There is, however, a serious volume problem with small diapirs. Diapirs with radii of less than 20 km and limited vertical extent (figure 6c) are not large enough to create the voluminous Hawaiian shields, typically 25000–50000 km^3 (Frey & Roden 1987). A possible alternative model (not shown in figure 6) is that solitary waves develop in a small radius plume conduit. Because these waves have closed streamlines, each wave can transport large amounts of geochemically distinct material from the source region (Whitehead & Helfrich 1990). Interaction of this conduit with the lithosphere might result in each shield forming from a different solitary wave.

An important aspect of the models in figure 6 is that the radius of the plume conduits are small, not more than 40 km, compared to the large, more than 1000 km topographic anomaly of the Hawaiian swell. However, this anomaly may result from horizontal divergence of the plume as it encounters the lithosphere (Loper 1991). Moreover, consideration of thermal-chemical plumes shows that the thermal effects are dispersed much more widely than the compositional effects (Liu & Chase 1991*b*).

6. Conclusions

The intershield geochemical differences between Hawaiian shields provide several important constraints on plume dynamics, especially the processes that convert a deep mantle plume into discrete shield volcanoes. Specifically, (i) adjacent shields must have long-lived, and distinct magma ascent paths and (ii) melt formation must be restricted to a zone less than 50 km in width. Also the plume is heterogeneous in isotopic ratios and incompatible element abundance ratios involving Ti, Sr and Nb. Because Rb/Sr and Sm/Nd are not correlated with ^{87}Sr/^{86}Sr and ^{143}Nd/^{144}Nd, source heterogeneity was created relatively recently or the melting process changed these trace element abundance ratios. The processes occurring during melt formation and migration probably control intershield differences in major element contents. Finally, interaction of the plume conduit with the lithosphere must enable adjacent shields to form from geochemically different sources and enable each shield to be

size and temperature, thereby leading to intershield differences in extent of melting and mean pressure of melt segregation. However, this model has a significant problem stemming from the size of the shields. A conservative estimate for large shield, like Mauna Loa, is 45000 km^3. If the mean extent of melting is not more than 20%, a small radius diapir is not large enough to form a volcanic shield.

characterized by a different mean extent of melting and pressure of melt segregation. Although the validity of these inferences requires further evaluation, it is evident that Hawaiian shields did not form from a geochemically homogeneous plume.

This research was supported by U.S. National Science Foundation Grants to F. A. F. and J. M. R. Data acquisition and graphics were facilitated by M. Chapman, W. McDonough, P. Dawson, T. Furman, P. Ila, J. Sparks, D. Tormey, and H.-J. Yang. We thank M. Feigenson and J. Sinton for allowing us to use their unpublished geochemical data for Waianae lavas, and D. Clague for a constructive review. Also our ideas have benefitted by discussions with I. Campbell, M. O. Garcia, D. Loper, R. Kinzler, M. Liu, D. Presnall, R. N. Thompson, S. Watson and J. Whitehead.

References

Budahn, J. R. & Schmitt, R. A. 1985 Petrogenetic modeling of Hawaiian tholeiitic basalts: a geochemical approach. *Geochim. cosmochim. Acta* **49**, 67–87.

Chen, C. Y. & Frey, F. A. 1985 Trace element and isotope geochemistry of lavas from Haleakala volcano, East Maui, Hawaii: Implications for the origin of Hawaiian basalts. *J geophys. Res.* **90**, 8743–8768.

Clague, D. A. & Dalrymple, G. B. 1987 The Hawaiian-Emperor volcanic chain. Part 1. Geologic evolution, USGS Prof. Pap. **1350**, 5–54.

Clague, D. A., Weber, W. S. & Dixon, J. E. 1991 Picritic glasses from Hawaii. *Nature, Lond.* **353**, 553–556.

Easton, R. M. 1987 Stratigraphy of Kilauea volcano. USGS Prof. Pap. **1350**, 243–260.

Eggins, S. M. 1992*a* Petrogenesis of Hawaiian tholeiites. 1. Phase equilibria constraints. *Contrib. Mineral. Petrol.* **110**, 387–397.

Eggins, S. M. 1992*b* Petrogenesis of Hawaiian tholeiites. 2. Aspects of dynamic melt segregation. *Contrib. Mineral. Petrol.* **110**, 398–410.

Eggler, D. H. 1978 The effect of CO_2 upon partial melting of peridotite in the system Na_2O–CaO–Al_2O_3–MgO–SiO_2–CO_2 to 35 kb, with an analysis of melting in a peridotite–H_2O–CO_2 system. *Am. J. Sci.* **278**, 305–343.

Frey, F. A. & Roden, M. F. 1987 The mantle source for the Hawaiian Islands: constraints from the lavas and ultramafic inclusions. In *Mantle metasomatism* (ed. M. A. Menzies & C. J. Hawkesworth), pp. 423–463. San Diego, CA: Academic Press.

Frey, F. A., Garcia, M. O. & Roden, M. F. 1993 Geochemical characteristics of Koolau Volcano: implications of intershield geochemical differences among Hawaiian volcanoes. *Geochim. cosmochim. Acta.* (In the press.)

Garcia, M. O., Muenow, D. W., Aggrey, K. E. & O'Neil, J. 1989 Major element, volatile and stable isotope geochemistry of Hawaiian submarine tholeiitic glasses. *J. geophys. Res.* **94**, 10525–10538.

Griffiths, R. W. & Campbell, I. H. 1991 On the dynamics of long-lived plume conduits in the convecting mantle. *Earth planet. Sci. Lett.* **103**, 214–227.

Griffiths, R. W. & Richards, M. A. 1989 The adjustment of mantle plumes to changes in plate motion. *Geophys. Res. Lett.* **16**, 437–440.

Hart, S. R., Hauri, E. H., Oschmann, L. A. & Whitehead, J. A. 1992 Mantle plumes and entrainment: isotopic evidence. *Science Wash.* **256**, 517–520.

Hofmann, A. W., Feigenson, M. D. & Raczek, I. 1984 Case studies on the origin of basalt. III. Petrogenesis of the Mauna Ulu eruption, Kilauea, 1969–1971. *Contrib. Mineral. Petrol.* **88**, 24–35.

Kinzler, R. J. & Grove, T. L. 1992 Primary magmas of mid-ocean ridge basalts. II. Applications. *J. geophys. Res.* **97**, 6907–6926.

Klein, E. M. & Langmuir, C. H. 1987 Global correlation of ocean ridge basalt chemistry with axial depth and crustal thickness. *J. geophys. Res.* **92**, 8089–8115.

Klein, F. W., Koyanagi, R. Y., Nakata, J. S. & Tanigawa, W. R. 1987 The seismicity of Kilauea's magma system. USGS Prof. Pap. **1350**, 1019–1102.

Kurz, M. D. & Kammer, D. P. 1991 Isotopic evolution of Mauna Loa volcano. *Earth Planet. Sci. Lett.* **103**, 257–269.

Langenheim, V. A. M. & Clague, D. A. 1987 The Hawaiian-Emperor Volcanic Chain. Part II. Stratigraphic framework of volcanic rocks of the Hawaiian Islands. *USGS Prof. Pap.* **1350**, 55–84.

Leeman, W. P., Budahn, J. R., Gerlach, D. C., Smith, D. R. & Powell, B. 1980 Origin of Hawaiian tholeiites: trace element constraints. *Am. J. Sci.* A**280**, 794–819.

Lipman, P. W., Rhodes, J. M. & Dalrymple, G. B. 1990 The Ninole basalt – implication for the structural evolution of Mauna Loa volcano, Hawaii. *Bull. Volcan.* **53**, 1–19.

Liu, M. & Chase, C. G. 1991*a* Evolution of Hawaiian basalts: a hotspot melting model. *Earth Planet. Sci. Lett.* **104**, 151–165.

Liu, M. & Chase, C. G. 1991*b* Boundary-layer model of mantle plumes of thermal and chemical diffusion and buoyancy. *Geophys. J. Int.* **104**, 443–440.

Liu, M., Yuen, D. A., Zhao, W. & Honda, S. 1991 Development of diapiric structures in upper mantle due to phase transitions. *Science, Wash.* **252**, 1836–1839.

Loper, D. E. 1991 Mantle plumes. *Tectonophys.* **187**, 373–384.

Loper, D. E. & Stacey, F. D. 1983 The dynamical and thermal structure of deep mantle plumes. *Phys. Earth planet. Into.* **33**, 304–317.

Loubet, M., Sassi, R. & Di Donato, G. 1988 Mantle heterogeneities: a combined isotope and trace element approach and evidence for recycled oceanic crust materials in some OIB sources. *Earth planet. Sci. Lett.* **89**, 299–315.

Macdonald, G. A. & Katsura, T. 1964 Chemical composition of Hawaiian lavas. *J. Petrol.* **5**, 82–133.

McKenzie, D. & O'Nions, R. K. 1991 Partial melt distributions from inversion of rare earth element concentrations. *J. Petrol.* **32**, 1021–1091.

Morgan, W. J. 1971 Convection plumes in the lower mantle. *Nature, Lond.* **230**, 42–43.

Nielsen, R. L. 1990 Simulation of igneous differentiation processes. In *Reviews in mineralogy*, vol. 24: *Modern methods of igneous petrology: understanding magmatic processes* (ed. J. Nichols & J. K. Russell), pp. 65–105. Washington, D.C.: Mineralogical Society of America.

Olson, P. 1990 Hot spots, swells and mantle plumes. In *Magma transport and storage* (ed. M. P. Ryan), pp. 33–51. Wiley.

O'Nions, K., Hamilton, P. J. & Evensen, N. M. 1977 Variations in $^{143}Nd/^{144}Nd$ and $^{87}Sr/^{86}Sr$ in oceanic basalts. *Earth planet. Sci. Lett.* **34**, 13–22.

Powers, H. A. 1955 Composition and origin of basaltic magma of the Hawaiian Island. *Geochim. cosmochim. Acta* **7**, 77–107.

Rhodes, J. M. 1983 Mantle heterogeneity associated with the Hawaiian melting anomaly: evidence from Mauna Loan lavas. *EOS, Wash.* **64**, 348.

Rhodes, J. M., Wenz, K. P., Neal, C. A., Sparks, J. W. & Lockwood, J. P. 1989 Geochemical evidence for invasion of Kilauea's plumbing system by Mauna Loa Magma. *Nature, Lond.* **337**, 257–260.

Ribe, N. M. 1988 Dynamical geochemistry of the Hawaiian plume. *Earth planet. Sci. Lett.* **88**, 37–46.

Ribe, N. M. & Smooke, M. O. 1987 A stagnation point flow model for melt extraction from a mantle plume. *J. geophys. Res.* **92**, 6437–6443.

Roden, M. F., Frey, F. A. & Clague, D. A. 1984 Geochemistry of tholeiitic and alkalic lavas from the Koolau Range, Oahu, Hawaii: implications for Hawaiian volcanism. *Earth planet. Sci. Lett.* **69**, 141–158.

Roden, M. F., Trull, T., Hart, S. R. & Frey, F. A. 1993 New He, Nd, Pb and Sr isotopic constraints on constitution of the Hawaiian plume: results from Koolau volcano, Oahu, Hawaii. *Geochim. cosmochim. Acta.* (In the press.)

Skilbeck, J. N. & Whitehead, J. A. 1978 Formation of discrete islands in linear island chains. *Nature, Lond.* **272**, 499–501.

Stearns, H. T. 1940 Four-phase volcanism in Hawaii (abstract). *Bull. geol. Soc. Am.* **51**, 1947–1948.

Stille, P., Unruh, D. M. & Tatsumoto, M. 1983 Pb, Sr, Nd and Hf isotopic evidence of multiple sources for Oahu, Hawaii basalts. *Nature, Lond.* **304**, 25–29.

Stille, P., Unruh, D. M. & Tatsumoto, M. 1986 Pb, Sr, Nd and Hf isotopic constrains on the origin of Hawaiian basalts and evidence for a unique mantle source. *Geochim. cosmochim. Acta* **50**, 2303–2319.

Tatsumoto, M. 1978 Isotopic composition of lead in oceanic basalt and its implication to mantle evolution. *Earth planet. Sci. Lett.* **38**, 63–87.

Thompson, R. N. 1987 Phase-equilibria constraints on the genesis and magmatic evolution of oceanic basalts. *Earth-Sci. Rev.* **24**, 161–210.

Tilling, R. I., Wright, T. L. & Millard, H. T. 1987 Trace-element chemistry of Kilauea and Mauna Loa lava in space and time: a reconnaissance. *USGS Prof. Pap.* **1350**, 641–689.

Watson, S. & McKenzie, D. 1991 Melt generation by plumes: a study of Hawaiian volcanism. *J. Petrol.* **32**, 501–537.

West, H. B., Gerlach, D. C., Leeman, W. P. & Garcia, M. O. 1987 Isotopic constraints on the origin of Hawaiian lavas from the Maui Volcanic Complex, Hawaii. *Nature, Lond.* **330**, 216–219.

White, W. & Hofmann, A. W. 1982 Sr and Nd isotope geochemistry of oceanic basalts and mantle evolution. *Nature, Lond.* **294**, 821–825.

Whitehead, J. A. & Helfrich, K. R. 1990 Magma waves and diapiric dynamics. In *Magma transport and storage* (ed. M. P.Ryan), pp. 53–76. Wiley.

Wilson, J. T. 1963 A possible origin of the Hawaiian Island. *Can. J. Phys.* **41**, 863–870.

Wright, T. L. 1971 Chemistry of Kilauea and Mauna Loa lava in space and time. *USGS Prof. Pap.* **735**, 40.

Wyllie, P. J. 1988 Solidus curves, mantle plumes and magma generation beneath Hawaii. *J. geophys. Res.* **93**, 4171–4181.

Zbinden, E. A. & Sinton, J. M. 1988 Dikes and petrology of Waianae Volcano, Oahu. *J. geophys. Res.* **93**, 14856–14866.

Melt production rates in mantle plumes

By Robert S. White

Bullard Laboratories, Madingley Road, Cambridge CB3 0EZ, U.K.

I calculate the melt production rates for mantle plumes lying beneath oceanic lithosphere from the crustal thickening measured by using seismics and from the volume of the overlying ridge. Observed melt production rates are higher where the lithosphere is thinner, in accord with theoretical predictions of the processes of decompression melting in convective plumes. The productivity of mantle plumes, and in particular that of the Hawaiian plume, is shown to vary on timescales of a few tens of millions of years. This can be explained by variations in the temperature and flow rate of the plumes. The trace of the Réunion plume shows a marked drop in melt production over the 30 Ma following generation of the Deccan flood basalts, which reflects a decrease in the plume temperature from the transient abnormally hot conditions associated with initiation of the plume.

1. Controls on decompression melting

Decompression melting of the mantle occurs within plumes where the mantle rises buoyantly until it is deflected by the overlying lithosphere, and beneath rifts where the thinning lithosphere allows the underlying mantle to well up. When rifting occurs above a mantle plume, extremely large volumes of melt may be generated rapidly, because abnormally hot mantle can decompress to shallow depths.

The rate of melt production is governed by the bulk composition of the mantle, the presence of volatiles or water, the temperature of the mantle, the lithospheric thickness, and the rate of decompression. Of these factors, the bulk composition of the mantle varies little, so does not cause variations in melt productivity. Though volatiles or water have a marked effect on melting in regions such as those above subduction zones, I assume that the melting beneath mid-ocean ridges and in mantle plumes is dry and use McKenzie & Bickle's (1988) method to calculate the volume of melt generated by decompression. I outline below the effect on melt production of variations in mantle temperature, lithospheric thickness, and decompression rate.

(a) Plume temperatures and shapes

Mantle plumes arise as boundary layer instabilities from deep in the Earth, though it is not yet clear whether they arrive at the surface directly from the core–mantle interface, or from the upper–lower mantle boundary. They are not necessarily axisymmetric, though are often modelled as such for computational reasons (Courtney & White 1986; Watson & McKenzie 1991). Rising spoke-like or triple junction patterns may be common at their base, particularly if there is significant internal heating (Parsons & Richter 1981; Houseman 1990). In places rising sheets of mantle may reach the surface, the 750 km long Cameroon Line and the 450 km long Rodrigues Ridge being possible examples.

Phil. Trans. R. Soc. Lond. A (1993) **342**, 137–153

© 1993 The Royal Society

Printed in Great Britain

137

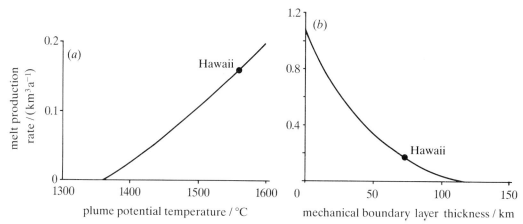

Figure 1. (a) Predicted variation of approximate melt production rate with potential temperature in the centre of the mantle plume; (b) predicted variation of melt production rate with thickness of the mechanical boundary layer. Both curves are from Watson & McKenzie's (1991) convection model for the present-day Hawaiian plume.

Plumes are not steady-state features. From time to time old plumes die out and new ones start. The mass flux and temperature of a new plume is higher than the subsequent flow, and many flood basalt provinces appear to be associated with the transient conditions accompanying the initiation of new plumes (White & McKenzie 1989). Many plumes have been documented as active over periods of the order of 100 Ma, but the difficulty of tracing them beneath continental lithosphere makes it uncertain whether any individual plumes have survived as long as 200 Ma.

The temperature within a plume can be estimated from the uplift and geoid anomaly it causes, by the increase in heat flow through the overlying seafloor, and by the rate of production and composition of the igneous rock it generates. Temperatures in mantle plumes are typically 250–300 °C above normal.

Watson & McKenzie (1991) show that although the average degree of melting beneath Hawaii is less than 7%, large volumes of melt are generated because the plume continually cycles mantle through the melting region. Figure 1a shows the approximate variation with temperature of melt production for models of the Hawaiian plume. A plume could exist beneath Hawaii with an excess temperature nearly 100 °C above the normal mantle temperature of 1280 °C, yet still produce negligible melt because the base of the lithosphere would prevent the convecting mantle rising sufficiently shallow to cross the solidus.

(b) Lithospheric thickness

Thick lithosphere prevents melting in plumes by limiting the decompression that can occur. In the Hoggar plume, for example, little melt is produced even though the plume causes regional uplift of about 1 km (Crough 1981). Conversely, thin lithosphere allows considerable melting in plumes.

Removal of melt reduces the density of the residual mantle (Oxburgh & Parmentier 1977; Bickle 1986). For slow-moving plates it is possible that depleted mantle accumulating above a plume may limit the vertical rise of the plume, thus reducing the amount of decompression melting that occurs. Conversely, heat conduction from a plume decreases the thickness of the overlying lithosphere, allowing increased decompression melting. The maximum lithospheric thinning from heat conduction is less than 10% (Courtney & White 1986; Watson & McKenzie 1991). Neither of these effects is large, and they work in opposite directions, so it is unlikely that they can be detected in the geological record.

(c) *Mantle plume flow rates*

Provided the lithosphere is sufficiently thin and the temperature sufficiently high for melting to occur, the melt production rate in plumes beneath intact plates is directly proportional to the mantle flow rate.

Beneath rifts away from hotspots, the mantle decompresses as it moves upward. For rapid stretching the heat loss by conduction can be ignored, and the total volume of melt generated by decompression depends only on the initial entropy and the amount of lithospheric thinning. At oceanic rifts, heat loss by conduction only causes a measurable decrease in the volume of melt spreading rates lower than 20 mm a^{-1} (White 1992).

Where a rift lies above a plume, as in many flood basalt provinces, the melt production depends on both the rate of stretching and on the mantle flow in the plume. At present these interactions are too complex to model satisfactorily. But it is likely that melting in many flood basalt provinces is controlled primarily by the rate of lithospheric rifting, because the melt production rates are so high (typically 5–10 km^3 a^{-1}) and the duration of the main igneous phase so short (typically 0.5–2 Ma), that they can be explained only by rapid and widespread decompression.

2. Melt distribution in the crust

Less than 1%, and possibly as little as 0.1%, of the melt remains in the source mantle region. The melt moving upwards is distributed as surface flows, as dykes and sills within the mid-crust, or as massive underplated or heavily intruded regions in the lower crust.

(a) *Surface flows*

Basaltic rocks can flow remarkably large distances. For example, subaerial flows in the Columbia River Basalt Group can be traced more than 750 km from the feeder dykes, with some flows exceeding 2000 km^3 and probably approaching 3000 km^3 in volume (Tolan *et al.* 1989). Underwater basaltic flows extend up to 60 km across a smooth sedimented surface in the northeast Pacific (Davis 1982). Where melt supply was unusually high, as in the Nauru Basin, western Pacific, huge flows similar to those found in subaerial flood basalts have been identified (Saunders 1985). They probably originated from the mantle plume volcanism presumed to have built the adjacent Ontong-Java Plateau (Tarduno *et al.* 1991; White *et al.* 1992).

Although flood basalts produce sufficient melt to feed enormous flows, both subaerial and submarine, the melt production rates from plumes beneath intact plates are much lower and individual eruptions are frequent and relatively small. The lava from oceanic volcanoes does not generally flow far underwater from the source. The flexural moat surrounding large oceanic volcanoes also restricts the outward movement of basaltic flows.

(b) *Dyke intrusion*

Lateral dyke propagation is controlled mainly by a balance between viscous and buoyancy forces in the molten rock, with the strength of the medium through which the dyke moves affecting only the structure near the advancing tip (Lister 1990; Lister & Kerr 1991). Provided the melt supply is sufficiently large, basaltic dykes can flow hundreds of kilometres laterally in continental crust. For example, the 1.27 Ga Mackenzie dykes extend 1500 km through the Archaean shield at mid-crustal levels. They were emplaced contemporaneously with the Akalulia and Coppermine River flood basalts, and appear to have been generated following rifting above a mantle plume in a similar manner to younger flood basalt provinces (LeCheminant & Heaman 1989). So we might expect similar dyking to occur in more recent flood basalt provinces, although it is often obscured by the basaltic cover. Only where this has been eroded off, can the dykes be readily seen.

Large dykes may propagate laterally less easily through oceanic than continental crust. Although it is almost impossible to determine whether dyke swarms are present in oceanic crust, the crustal thickening beneath hotspot islands appears to be localized above the plume. This suggests that there has been little lateral melt transportation, in contrast to continental settings where long-term uplift beneath flood basalts provides evidence of considerable crustal thickening and lateral melt movement (McKenzie 1984; Cox 1992). Around the Hawaiian plume, for example, Lindwall (1988) shows from seismics that the crust as close as 300 km to Hawaii is no thicker than normal oceanic crust, and Helmberger & Morris (1970) demonstrate the same on a profile 350 km from Hawaii. Seismic and gravity profiles across other volcanic ridges give similar results: there are abrupt decreases in crustal thickness on the margin of Iceland (Gebrande *et al.* 1980; Ritzert & Jacoby 1985); across the edge of the Madagascar Ridge (Sinha *et al.* 1981); and across the flanks of the Chagos-Laccadive Ridge (Francis & Shor 1966).

Dykes may propagate less easily through oceanic than through continental crust because the basaltic melt generated in mantle plumes is similar in composition to that of the oceanic crust through which it intrudes, and so has a similar density. Furthermore, oceanic crust is only about 20% the thickness of continental crust. Both factors mean that the level of neutral buoyancy is shallower in oceanic than in continental crust and much of the melt probably extrudes before it has the opportunity to migrate significant distances laterally.

(c) *Underplating and lower crustal intrusion*

The evolved, uniform composition of tholeiitic flood basalts and of the majority of ocean island volcanics suggests that fractionation within or below the crust has left considerable volumes of residual igneous rock at depth (Cox 1980, 1992). When solidified in the lower crust the primary melts from plumes produce rocks with high seismic velocities and densities. Such rocks are found beneath volcanic rifted continental margins (White & McKenzie 1989; White 1992), and beneath intra-plate volcanic ridges (Sinha *et al.* 1981; Watts *et al.* 1985*a*; Zucca *et al.* 1982; ten Brink & Brocher 1987). Sometimes the primary lavas reach the surface without fractionating, as do the extensive picrites of the Lebombo monocline in the Karoo Province. However, more commonly they are trapped in the lower crust by their greater density, and can only rise to the surface when the density of the melt has been lowered by fractionation of olivine crystals (Stolper & Walker 1980). This explains both the relatively uniform composition of most tholeiitic basalts in flood basalts and ocean islands, and also the presence of large underplated igneous volumes.

The percentage of the melt emplaced in the lower continental crust is much greater than in oceanic crust. The main reason for this is probably the reduced density of continental crust, which makes it harder for the melts to rise through it than through the higher density oceanic crust. Furthermore, the melting temperature of the continental crust may be as low as 600 °C, so the large quantities of melt at temperatures greater than 1200 °C that are injected into continental crust rapidly melt it and thus generate an even more effective density trap. The density and melting temperatures of oceanic crust are higher than those for continental crust, so it does not form such a good density trap.

Seismic studies suggest that 60–80 % of the melt is underplated on volcanic continental margins, but that much less is underplated beneath ocean islands. Under the Hawaiian Islands, for example, Watts *et al.* (1985*b*) and ten Brink & Brocher (1988) suggest from seismics that only about 40 % of the igneous melt is underplated, while Lindwall (1988) re-interpreted one of the same profiles to suggest that less than 15 % of the melt is underplated. In part the different estimates for the Hawaiian Islands highlight the ambiguity in interpreting lower crustal velocities, as either original oceanic crust or as newly intruded igneous rock. Nevertheless it is clear that a far smaller proportion of the melt is underplated beneath oceanic crust than beneath continental crust.

3. Melt production rates in mantle plumes

The melt emplacement rate in the crust varies on several timescales. From days to hundreds of years, the rates reflect processes at crustal levels as magma chambers inflate and are emptied. Up to a few millions of years the rates are governed largely by the lithosphere as melt rises through it from the underlying mantle. Many hotspots produce chains of individual islands or seamounts as melt is focused into one or more volcanoes, before jumping on to a new location. For example, over 100 separate volcanic centres exist along the 5700 km long Hawaiian-Emperor chain (Bargar & Jackson 1974). This focusing is probably caused by the rigid lithosphere and the necessity to open conduits through it for the passage of melt.

I am here concerned with longer term changes in the behaviour of mantle plumes. By averaging the rates measured at crustal levels over periods of about 10 Ma, I smooth out the fluctuations caused by solitary waves in the mantle, or by the control exerted by the lithosphere and crust.

(a) *Estimation of melt production rates*

Melt production rates are best estimated by using seismics to define the crustal thickness. I here consider plumes beneath oceanic plates because the thickness of the pre-existing crust is well known and varies little from 7 km away from obvious tectonic structures such as fracture zones (White *et al.* 1992). The measured crustal thickening can therefore be attributed to new igneous addition, and includes both extrusive and intrusive components.

The only seismic profiles across volcanic ridges modelled with synthetic seismograms are from the Hawaiian Ridge (Lindwall 1988), the Madagascar Ridge (Sinha 1981) and Kerguelen Plateau (Recq *et al.* 1990). However, older profiles interpreted using slope-intercept solutions are more widespread and still give an indication of the crustal thickness provided allowance is made for the likelihood that the crustal thicknesses are about 20 % too small because of the assumption of plane layers rather than the more likely velocity gradients (White *et al.* 1992; see also plane-layer versus synthetic seismogram interpretations of the Madagascar Ridge in Sinha *et al.* (1981)).

Even using results only from volcanic chains with seismic control, we must still extrapolate from the few locations where the thicknesses are known. For this, I assume local Airy compensation and estimate the total igneous thickness from the volume of the ridge above a base level, calculated by digitizing bathymetric contours. This is similar to Schubert & Sandwell's (1989) approach, except that they used $15' \times 15'$ bathymetric averages and a different method for defining the excess volume of the volcanic ridges.

There are two major uncertainties in calculating the excess volume of a ridge, both of which arise from the difficulty of defining the base level. An active plume causes up to 1–2 km uplift and affects a region of up to 2000 km diameter. After a plume has moved away, it takes 50–100 Ma for the excess heat to conduct through the lithosphere, and for the bathymetry to subside to its pre-plume depth. Superimposed on this is permanent uplift caused by the decreased density of partly depleted mantle, which remains even after the plume has gone. Both mechanisms elevate the seafloor around volcanic ridges above the normal oceanic subsidence curve.

I here define the boundary of a volcanic ridge as the location of the break in slope from the gentle, long-wavelength swell, to the steep volcanic edifice. Where the lower part of the ridge has been buried in thick sediments, I extrapolate downward to a deeper base level representing the top of the igneous basement rather than using the present seafloor.

The total igneous volume is calculated assuming Airy isostasy, with a mean density difference of 0.3 Mg m^{-3} between the crust and mantle. Several approximations are involved in this. The density of the underplated region is probably greater than that of the extruded volcanics, which would lead to a thicker crust than I calculate. However, the density of the depleted mantle is probably lower beneath the ridge than elsewhere, and if this were properly included, I would calculate a thinner crust. Since the densities are not well known, my simple assumption of a single density difference is reasonable, and is in any case probably not far wrong because these two sources of error work in opposite directions. Where the volcanic ridge was constructed on very young lithosphere, the assumption of local Airy isostasy is better than where the ridge was emplaced on older lithosphere and part of the load is supported over a wide area (e.g. Hawaiian Ridge (Watts & ten Brink 1989)). Nevertheless, I have confidence that this straightforward approach gives useful answers by ensuring that the total igneous volume predicted by this method is in agreement with the total volume determined from seismic measurements, at the locations where both methods are available.

(b) Melt variation with time: the Hawaiian plume

For the past 35 Ma the Hawaiian plume has lain beneath oceanic lithosphere some 80 Ma older than itself (figure 2, top). This removes lithospheric thickness variations as a possible control on the melt production rate and leaves variations in the temperature and flow rate of the plume itself as the major variable factor. The present melt production rate measured from seismics near Oahu (Watts *et al.* 1985*b*) is 0.14–0.18 km^3 a^{-1} (square symbols, figure 2), in agreement with the 10 Ma running average calculated assuming Airy isostasy.

Over the past 35 Ma the melt production of the Hawaiian plume has increased steadily by an order of magnitude. If the plume flow rate has remained unchanged over that period, a temperature increase of about 150° at the centre would explain this increase (figure 2). Alternatively, if the flow rate of the plume has increased over the past 35 Ma then a smaller change in plume temperature would be required to

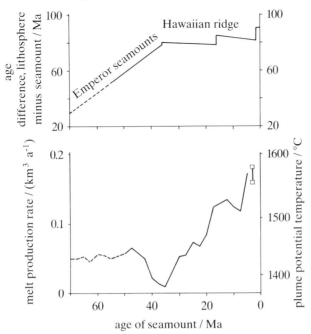

Figure 2. Lower diagram shows variation of melt production rate (left-hand scale) with age along the Hawaiian-Emperor seamount chain. Total melt rate is calculated from the volumes above the break in slope at the seafloor of individual shield volcanoes (Bargar & Jackson 1974), extrapolated to include underplated region by assuming local Airy isostasy. A running average of 10 Ma has been calculated every 2.5 Ma along the seamount chain, omitting the currently active volcanoes of Mauna Loa, Kilauea and Loihi. Open squares show the range in mean igneous production rate from a cross-section based on seismics near Oahu from Watts *et al.* (1985*b*), with the range indicating different assumptions of how much igneous material has been eroded off into the moat adjacent to the islands. Right-hand scale shows approximate variations in the potential temperature at the centre of the plume which would account for the variations in melt production rate over the period 35–0 Ma, assuming that the flow in the mantle plume remained unchanged from that deduced by Watson & McKenzie (1991) for the present day (see figure 1*a*). Top diagram shows age of the lithosphere at the time of emplacement of seamounts along the Hawaiian-Emperor chain (after Clague & Dalrymple 1987). During 35–0 Ma the plume has remained beneath lithosphere consistently 80 Ma older than the seamounts, while before 35 Ma, the older the seamounts, the younger the lithosphere at the time of emplacement.

explain the increased melt production. In either case it is clear that the vigour of the mantle plume has increased over a timescale of 35 Ma, making Hawaii at the present day the most active intra-plate volcano in the world.

Over the period 50–35 Ma the melt rate was decreasing to the low at 35 Ma (figure 2). At 50 Ma the lithosphere on which the seamounts were being emplaced was some 20 Ma younger than at 35 Ma (figure 2, top), but the concomitant increase in lithospheric thickness from 50–35 Ma is too small to explain most of the melt production rate drop over the same period.

Melt production was approximately constant during the period 70–50 Ma. However, this oldest section of the Emperor Seamounts is also partly buried by terrigenous sediments so it is likely that the total volumes of the seamounts are underestimated. Correction for this would increase the melt production rates along the oldest remaining section of the volcanic chain.

I conclude that changes in the plume temperature and flow, over timescales of a few tens of Ma are responsible for the long-term changes in melt production recorded along the Emperor-Hawaiian seamount chain.

(c) Melt variation with time in the Réunion plume

The Réunion plume spent the first 30 Ma of its recorded history beneath very young lithosphere, so it provides a contrast to the Hawaiian plume. From the time of the Deccan flood basalts at 66 Ma, probably coinciding with initiation of the Réunion plume, until about 35 Ma when it crossed the Central Indian Ridge, the plume lay close to the spreading ridge (Schlich 1982; Royer *et al.* 1989) and generated the huge volcanic ridge beneath the Laccadives, the Maldives, Chagos Bank and Saya da Malha Bank. Volcanic dates from borehole samples show a progression along the chain, with southern Chagos Bank and Saya da Malha Bank being formed at the same time, but subsequently split by continued seafloor spreading along the Central Indian Ridge (Duncan & Hargraves 1990).

Airy isostatic balances of the excess ridge topography indicate crustal thickening of about 20 km, in accord with sparse seismic refraction measurements reported by Francis & Shor (1966). Their reversed refraction line 4–5 in a channel between the Maldive and Chagos archipelagos did not reach Moho, despite a 100 km line length, from which they conclude that the Moho is unlikely to be shallower than 20 km. They also recorded a mid-crustal refractor with a velocity higher than 7.1 km s^{-1}, which is typical of the underplating found elsewhere under volcanic ridges. Further north, a profile between the Laccadives and Maldives recorded Moho at just over 17 km depth: taking account of velocity gradients and the probably higher velocity of material in the lower crust would increase this to well over 20 km.

Melt production over 66–35 Ma shows a steady decrease from the peak rates in the Deccan (figure 3). These rates are an order of magnitude higher than those of the Hawaiian-Emperor chain, reflecting the thinner lithosphere and the hotter mantle in the initiating plume. The Indian plate was moving northwards rapidly with respect to the Réunion plume during this period, allowing the melt to migrate upward to generate a thick igneous ridge, which reached well above sea-level. Since the plume lay under young lithosphere during the period 64–35 Ma, the decrease in melt production is probably due to decreasing temperature and flow rate in the plume following the peak as it initiated. The timescale of several tens of millions of years over which this occurred is similar to the timescale over which the Hawaiian plume changes have taken place.

As the plume passed beneath older lithosphere south of the Central Indian Ridge, the melt production continued to decrease, building first a series of smaller ridges, including Nazareth and Cargados-Carajos Banks, and then individual volcanic shields like Mauritius and Réunion, similar to the isolated seamounts generated along the Hawaiian-Emperor chain. Correction of the melt production rates to those that would have been generated under 20 km thick lithosphere comparable to that beneath the Chagos-Laccadive Ridge, flattens out the rate of decline in the melting (open symbols, figure 3). The vigour of the present plume is now comparable to that of mature plumes elsewhere.

(d) Melt variation with lithosphere thickness

Variations in melting caused by changes in the plume temperature are superimposed on the control exerted by the lithospheric thickness. The major plumes show a decrease in melt production rates with increasing lithospheric thickness

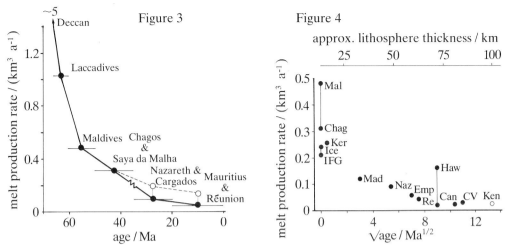

Figure 3. Variation of melt production rate with age along the trace of the Réunion plume, calculated from the bathymetric expression of the volcanic ridge, assuming Airy isostatic equilibrium and with thickness control from seismic refraction experiments of Francis & Shor (1966). Volcanic ages are from Duncan & Hargraves (1990), with horizontal bars showing period over which the rate is averaged. From the time of the Deccan flood basalts at 66 Ma until approximately 35 Ma when the plume crossed the spreading axis of the Central Indian Ridge, the plume remained beneath young lithosphere. Since 35 Ma the plume has migrated beneath progressively older lithosphere with Réunion island currently being built on 67 Ma oceanic crust. Open circles are an approximate correction of the measured rates back to those that would be expected beneath 20 km thick lithosphere to make it comparable to the rates from 66–35 Ma, using the melt rate against lithosphere thickness curve shown in figure 1b.

Figure 4. Melt production rates in mantle plumes (see table 1 for key) against the square root of the age of the oceanic lithosphere beneath which they lay. Approximate lithospheric thickness at the time of emplacement is shown along the top. The Kenyan plume (open symbol) lies beneath continental lithosphere about 100 km thick (see text for details). Note that for any given lithospheric thickness, variations in the plume temperature or in the vigour of mantle flow may cause variations in the melt production rate, such as are indicated for the Hawaiian Chain, for Iceland and the Iceland–Faeroes–Greenland Ridge, and for the Chagos and Maldive Ridges. Compare the general trend in observed rates with those predicted from convection modelling in figure 1b.

(figure 4 and table 1), similar to that predicted by simple convection models of plumes (figure 1b). Where the variations with age are known, as in the cases of the Hawaiian and Réunion plumes, I show the range on figure 4. Elsewhere I average across several tens of Ma to obtain a long-term mean.

(i) *The Iceland plume*

Igneous activity on the rifted continental margins of the northern North Atlantic shows a peak at break-up, caused by the transient abnormally hot mantle in the newly initiated Iceland plume. Igneous thickness estimates on the rifted margin and in the adjacent oceanic basin suggest that the mantle temperature dropped by more than 50 °C during the first 10 Ma following arrival of the new plume (Morgan *et al.* 1989; White 1992).

Oceanic crust adjacent to the Iceland plume averages 10.3±1.7 km thick (White 1992). Minor variations in melt productivity are suggested by V-shaped discontinuities that propagate across the Irminger and Iceland Basins, and by the appearance of seaward-dipping reflectors on the Kolbeinsy Ridge between 20–10 Ma.

Table 1. *Melt production rates above mantle plumes*

(Table shows melt production rates in excess of the 7 km of melt which generates normal thickness oceanic crust. The age of plume volcanism is the range used to calculate the melt productivity rate shown in the last column.)

abbreviation on figure 4	plume trace	age of plume volcanism/Ma	melt rate $\overline{km^3\,a^{-1}}$
Can	Canaries	60–0	0.02
Chag	Chagos plus Saya-da-Malha Banks	50–35	0.31
CV	Cape Verdes	25–0	0.03
Emp	Emperor Seamounts	60–40	0.06
Haw	Hawaiian Ridge	40–0	0.03–0.16
Ice	Iceland	35–0	0.24
IFG	Iceland–Faeroes–Greenland Ridges	55–35	0.21
Ken	Kenyan Rift	23–0	0.02
Ker	North Kerguelen plus Broken Ridge plus 90E Ridge	80–38	0.25
Lacc	Laccadive Ridge	66–60.5	1.03
Mad	Northern Madagascar Ridge	120–95	0.12
Mal	Maldives Ridge	60.5–50	0.48
Naz	Nazareth Ridge plus Cargados–Carajos Bank	35–20	0.09
Re	Réunion and Mauritius	20–0	0.04

The Iceland–Faeroe Ridge generated directly above the plume is 30–35 km thick (Bott & Gunnarsson 1980). Assuming Airy isostasy, and with an allowance for an equal volume of igneous rocks in the largely unexplored Iceland–Greenland Ridge, this suggests an excess melt production rate above that in the adjacent oceanic spreading centre of 0.21 km^3 a^{-1} during 55–35 Ma. However, as Bott & Gunnarsson (1980) comment, this thickness estimate is too small if there are high velocities or velocity gradients in the lower crust. The most probable rate, after allowance for the likelihood of increased velocities in the lower crust, is about 0.26 km^3 a^{-1}. There is some indication that the crust is thinner towards the younger, northwestern end, which would indicate gradually decreasing temperatures with time.

Beneath Iceland itself, the crustal thickness is uncertain because normal mantle velocities are not reached even by a depth of 50 km. Seismic velocities increase from about 7.0 km s^{-1} at 15 km depth to 7.6 km s^{-1} at 50 km depth, with a band of higher velocity layers at about 30 km (Gebrande *et al.* 1980). These velocities and the absence of a mature Moho are similar to the structure found beneath many active oceanic spreading centres, with the most likely explanation being that they are caused by enhanced temperatures and the presence of melt. The maximum crustal thickness is probably about 30 km, corresponding to the higher velocity lenses at that depth, and the minimum crustal thickness is around 20–24 km, corresponding to a reflector recorded at that depth in the south of Iceland (Bjarnason *et al.* 1992). With the present extent of the rift beneath Iceland and the full spreading rate of 19 mm a^{-1}, this gives an excess melt production rate of 0.12–0.24 km^3 a^{-1} above that beneath the adjacent Reykjanes Ridges caused by passive upwelling. I favour the higher rate of 0.24 km^3 a^{-1}, which is in agreement with the excess volume for the past 35 Ma calculated assuming Airy compensation by Schubert & Sandwell (1989), because this agrees well with the rate from the Iceland–Faeroes Ridge and there is no evidence from crustal thicknesses on the adjacent oceanic crust of a marked decrease in mantle temperatures over the past 50 Ma.

(ii) *Madagascar Rise*

The Madagascar Rise extends 1300 km southwards from Madagascar. It marks the trace of a plume which was responsible for the Upper Cretaceous Madagascar flood basalts formed as Gondwana broke up (Mahoney *et al.* 1991), and which generated the recent volcanism of Prince Edwards and Marion Islands, and possibly also of Crozet Island. The northern Madagascar Rise, with a water depth of 2–3 km, and 0.5–1.0 km of undisturbed sediments, was formed as the plume lay beneath newly created oceanic lithosphere perhaps 10 Ma old. Subsequently, the plume crossed the Southwest Indian Ridge, and its track thereafter is uncertain. Melt production increased as the plume crossed the zero-age lithosphere beneath the spreading centre, as is indicated by the thicker ridge in the southern part of the Madagascar Rise, topped by Walters Shoal which reaches to within 20 m of the sea surface.

I here consider melt production in just the northern Madagascar Rise, where the crustal structure is well constrained by seismics modelled by using synthetic seismograms (Sinha *et al.* 1981). These show a 20–25 km thick crust, with an 8 km thick lower crustal layer of about 7.6 km s^{-1}, representing the underplated more mafic region. Plume production rates calculated from both the seismic cross section and from gravity modelling assuming Airy compensation are about 0.12 km^3 a^{-1}.

(iii) *Canary plume*

For the past 60 Ma the plume which now lies beneath the Canary Islands has traversed beneath Mesozoic oceanic lithospheric in the eastern Central Atlantic some 110 Ma older than the plume. Holik & Rabinowitz (1992) interpret a 150–200 km wide zone of chaotic reflectors within the Cretaceous sediments as extrusive volcanics averaging just over 1 km thick. In the same area at the base of the crust they report a 200 km wide, 1.0–3.5 km thick high-velocity (7.1–7.4 km s^{-1}) layer interpreted as underplated igneous rocks. These igneous rocks added to the crust lie along the track of the Canary plume and coincide with elevated subsidence curves consistent with the uplift caused by a plume. The melt production rate during the period 60–50 Ma calculated from the inferred igneous thicknesses is 0.01–0.02 km^3 a^{-1}.

The Canary Islands themselves represent the most recent activity of the plume, with ages decreasing irregularly from the east to the west, consistent with the motion of the plate. Many of the islands exhibit a long (more than 20 Ma) magmatic history, a result of the slow motion of the plate. They are composed dominantly of shield-building basalts with local tholeiites, a consequence of the thick lithosphere and the resulting small degrees of mantle melting. Seismic refraction results from beneath the islands have not been modelled with synthetic seismograms, but suggest that the Moho is at least 17 km deep. A 3 km thick basal layer with a velocity of 7.1–7.5 km s^{-1} (Bosshard & Macfarlane 1970) probably represents underplated igneous rocks. Basal velocities of 7.4 km s^{-1} beneath Lanzarote and Fuerteventura were interpreted by Banda *et al.* (1981) as mantle, but may also represent intruded igneous rocks. From seismic and gravity data the average rate of accreting underplated igneous rock is 0.01 km^3 a^{-1}, and the rate of extrusion in island building is a little under 0.01 km^3 a^{-1} (Schmincke 1982), giving a total melt production rate of 0.02 km^3 a^{-1}.

(iv) *Cape Verde plume*

Like the Canaries, the Cape Verde islands were generated by a plume beneath late Jurassic oceanic lithosphere. The main igneous activity has occurred over the past 25 Ma, but the slow relative motion of only 9 mm a^{-1} between the African plate and the plume means that there is no clear age progression in the islands, other than a general younging towards the west, and that individual islands have been active for long periods. The 8 m geoid anomaly, 2 km uplift and 25% increase in heat flow beneath the islands have been modelled by convection in a mantle plume (Courtney & White 1986). Seismic refraction profiles yield an average crustal thickness of 11.4 km beneath the archipelago (Dash *et al.* 1976). Since these were interpreted by using the slope-intercept method, the true crustal thickness is probably about 20% greater or approximately 14 km. Across the archipelago, this yields a melt production rate in the mantle plume of 0.03 km^3 a^{-1}, similar to that found for the Canaries.

(v) *Kenya plume*

The plume beneath Kenya has produced regional uplift (Ebinger *et al.* 1989) and volcanism in the Kenyan Rift over the past 30 Ma. Volcanism has been most active since the Miocene and I here consider only melt production rates during the last 23 Ma. The total igneous volume has to be estimated from the mapped volumes of extrusive basalt by allowing for fractional crystallization and calculating the amount of gabbro that is likely to be left in the lower crust: the volume of intrusives found in this way account for about 70% of the total igneous rock, which is consistent with the ratio of intrusive to extrusive rocks observed elsewhere in continental crust by using seismic methods. Excluding rhyolites, which are presumed to be crustal melts and therefore not directly from the mantle plume, and excluding hyperalkalic rocks in a volcanic chain to the west of the Kenyan Rift, 212000 km^3 of melt was produced in the Miocene, 166800 km^3 in the Pliocene and 38400 km^3 in the Quaternary (Karson & Curtis 1989, and sources referenced therein). These yield an overall average melt production rate of just under 0.02 km^3 a^{-1}, with a similar rate of 0.02 km^3 a^{-1} in the Miocene and the Pliocene considered separately. There are indications of shorter-term variations in melting, with the rate dropping to about half in the last 2.5 Ma represented by the Quaternary, but being perhaps twice as high in the last half of the Pliocene (Williams 1972).

It is not easy to define the lithosphere thickness beneath the Kenyan Rift, although seismic reflection (Karson & Curtis 1989) and wide-angle refraction (Green *et al.* 1992; KRISP Working Party 1991) data indicate that crustal extension and thinning, and therefore by implication lithospheric thinning, has occurred. Karson & Curtis (1989) suggest that crustal extension has been less than 10% across the Kenyan Rift, and that much of the crustal thinning has been compensated by igneous intrusion from the plume. KRISP (1991) show that variable crustal thinning reaching 10–15 km has occurred within parts of the Kenyan Rift, and suggest that extension by between 10 and 35–40 km has occurred across different parts of the rift. Given the present rift valley width, this suggests local extension across the rift reaching 10–50%. If the original lithospheric thickness was 150 km, and stretching was by pure shear, then the lithosphere would be thinned beneath the rift to a minimum of about 100 km, so I take this as a representative thickness for plotting the Kenyan plume melt rate on figure 4. Note that Green *et al.* (1992) report the lithosphere–asthenosphere boundary as rising to depths of only 35–65 km beneath the rift, but they use a definition for asthenosphere as mantle containing partial melt (which causes reduced seismic velocities). In considering the limits on decompression of rising material in a mantle plume it is the mechanically rigid layer which forms the lithospheric lid, irrespective of how much partial melt may be migrating through it, so Green *et al.*'s definition of very shallow asthenosphere is not appropriate for our purposes.

(vi) *Other plumes*

I have only shown on figure 4 results from those plumes where there is reasonably good control on the melt production rates. However, there are many other plumes where it is clear that the same dependence is found for melting as a function of lithospheric thickness. For example, the production rate from the Hoggar plume beneath unbroken, thick continental lithosphere is extremely low, while the nearby Afar plume beneath a major continental rift has produced flood basalts and a thick underplated region (White & McKenzie 1989).

In the South Atlantic, the Tristan plume lay beneath the spreading axis from 120–70 Ma creating the Rio Grande Rise and the eastern portion of the Walvis Ridge with crustal thickening of 8–15 km deduced from gravity measurements (Detrick & Watts 1979). From 70 Ma onwards the plume lay away from the spreading axis, and created a volcanic ridge on just the African plate, with reduced crustal thickening of 7–10 km. At present, Tristan da Cunha is some 450 km off the axis with much reduced melt productivity. So in this case, also, a progressive decrease in melting accompanies migration of the plume beneath thicker lithosphere.

There are also many other smaller plumes, such as the Saint Helena and Ascension plumes, which probably have lower temperatures or reduced mantle flow rates than the major plumes considered in this section, and which would therefore plot below the points shown on figure 4.

(e) *Melt variations with plate velocity*

There is a hint from my results that melt productivity is higher from mantle plumes beneath fast-moving plates than from those beneath slow-moving plates, although the sample size is insufficient to separate out conclusively this possible effect from temporal changes in mantle flow. For plumes beneath the ridge axes of slow-spreading oceans, the production of large quantities of melt itself thickens the overlying lithosphere by freezing to form new crust and hence serves to reduce further melt generation by reducing the amount of decompression possible. Furthermore, the reduced density of the residual mantle left after extraction of partial melt also tends to inhibit decompression of the underlying mantle plume on slow-spreading ridges.

In contrast, where the plume lies beneath a fast-moving plate, the plate continually sweeps away the thickened crust and residual mantle. In Watson & McKenzie's (1991) convection model of the Hawaiian plume the radial velocity of the asthenospheric mantle is about 70 mm a^{-1} at a distance of 200 km from the plume centre, reducing to 40 mm a^{-1} at 400 km distance. Since the Pacific plate is moving across the Hawaiian plume at a rate of about 100 mm a^{-1} (Gripp & Gordon 1990) the plate velocity is higher than the horizontal outward flow of mantle from the plume so the plume moves progressively beneath fresh lithosphere. Although the Canary, Cape Verde and Réunion plumes lie beneath oceanic lithosphere of similar age to that beneath Hawaii, the overlying African plate is moving much more slowly than the

Pacific plate, at only 9–13 mm a^{-1} (Gripp & Gordon 1990), and the melt production rates are much lower (figure 4). For the plumes beneath the African plate the horizontal mantle flow rates fed by the plume are always much higher than the plate velocity, so the centre of the plume where decompression melting occurs always lies beneath a region which has been thoroughly affected by radial flow from the plume for a long period before transit across the centre of the plume.

Similarly reduced melt productivity is apparent in plumes beneath very young slow-spreading ridges. The Chagos–Maldive–Laccadive Ridge was formed on very young oceanic crust, as was the Faeroe–Iceland–Greenland Ridge. But melt production rates were much greater in the Chagos–Maldive–Laccadive Ridge where the plates were moving rapidly (120 mm a^{-1}) than in the Faeroes–Iceland–Greenland Ridge where the plates were moving only slowly (20 mm a^{-1}).

(f) Melt production in plumes beneath continental rifts

Where plumes lie beneath continental rifts, high melt production rates occur, particularly if the rifting occurs during the transient thermal conditions accompanying initiation of a new plume (White & McKenzie 1989). The relationship between rifting and magmatism is complex: the presence of a plume provides significant uplift which fosters extension and rifting, but huge volumes of melt can only be generated very rapidly if the mantle can decompress to shallow depths beneath stretched and thinned lithosphere. The earliest igneous activity is likely to occur as soon as the plume impinges on the lithosphere, and before major rifting occurs, but the volumes of such melts are likely to be small. On the other hand, the bulk of the melt generated by mantle decompression will be produced before the onset of seafloor spreading proper, by which time the continental lithosphere has rifted fully. So if the oldest seafloor spreading anomaly or the date of opening of the new ocean is used as the age of rifting, then it will post-date the majority of the flood basalts. Unravelling this complex interaction between tectonics and magmatism requires better age control than exists at present in many flood basalt provinces.

It is, however, clear that massive outbursts of igneous activity can only result if there is thin lithosphere above a mantle plume, and preferably a newly initiated plume for maximum effect. The thin lithosphere may result from continental break-up, perhaps itself encouraged by the presence of a plume, or as in the case of the Ontong–Java Plateau, may result from the arrival of a plume beneath very young, and therefore thin oceanic lithosphere.

4. Conclusions

Magmatism resulting from mantle plumes is controlled primarily by the overlying lithospheric thickness. If the lithosphere is thick, little melt will be produced because insufficient decompression can occur. Where the lithosphere is thin, either because the plume lies beneath young oceanic lithosphere or beneath rifting continental or oceanic lithosphere, then large quantities of melt are produced by mantle decompression. The transient conditions accompanying the initiation of new plumes produce particularly large melt productivity. Variations in the temperature and flow rates of mature mantle plumes occur over timescales of several tens of Ma.

The rate at which the overlying plate moves with respect to a plume also appears to exert some control on the melt productivity. More melt is generated in plumes beneath fast-moving plates than in those beneath slow-moving plates, probably because the fast-moving plate clears away the products of melting and the residual depleted mantle leaving the plume beneath pristine lithosphere.

Once melt has been generated in a plume it bleeds rapidly upward to the crust. It may then flow considerable distances laterally away from the source region, either as surface flows or as dykes at mid-crustal levels, provided the melt supply rates are sufficiently high to allow flow without freezing. Considerably more melt is underplated and travels laterally as sills or as dykes in continental crust than in oceanic crust. At the distal ends of dykes the melt may intrude at shallow levels or may feed local igneous centres. Therefore the final distribution of igneous rocks, and their history of fractionation and of crustal assimilation is likely to be far more complex than might be suggested by the relatively straightforward process of their generation by decompression melting in a plume or beneath a rift. A major key to understanding these relationships is good dating of the absolute ages and the rates of magmatism and of tectonic processes. Allied to these physical constraints, the geochemistry of the melts themselves provide evidence of the conditions under which the melt was generated (McKenzie & O'Nions 1991; Ellam 1991; White *et al.* 1992). The combination of knowledge of the rates and volumes of melt production with the geochemistry and petrology of the melts themselves is a powerful tool in understanding processes of melt generation in mantle plumes.

This work forms part of a study of the continental and oceanic lithosphere supported in part by the Natural Environment Research Council. Contribution number ES2849.

References

Banda, E., Danobeitia, J. J., Surinach, E. & Ansorge, J. 1981 Features of crustal structure under the Canary Islands. *Earth planet. Sci. Lett.* **55**, 11–24.

Bargar, K. E. & Jackson, E. D. 1974 Calculated volumes of individual shield volcanoes along the Hawaiian-Emperor chain. *J. Res. U.S. Geol. Survey* **2**, 545–550.

Bickle, M. J. 1986 Implications of melting for stabilisation of the lithosphere and heat loss in the Archaean. *Earth planet. Sci. Lett.* **80**, 314–324.

Bjarnason, I. Th., Menke, W., Florenz, O. G. & Caress, D. 1992 Tomographic image of the spreading center in southern Iceland. *J. geophys. Res.* (Submitted.)

Bosshard, E. & Macfarlane, D. J. 1970 Crustal structure of the western Canary Islands from seismic refraction and gravity data. *J. geophys. Res.* **75**, 4901–4918.

Bott, M. H. P. & Gunnarsson, K. 1980 Crustal structure of the Iceland–Faeroe Ridge. *J. geophys.* **47**, 221–227.

Campbell, S. M. & Griffiths, R. W. 1990 Implications of mantle plumes for the evolution of flood basalts. *Earth planet. Sci. Lett.* **99**, 79–93.

Clague, D. A. & Dalrymple, G. B. 1987 The Hawaiian-Emperor volcanic chain. I. Geologic Evolution. In *Volcanism in Hawaii* (ed. R. W. Decker, T. L. Wright & P. M. Stauffer), pp. 5–50. US. Geological Survey Professional Paper 1350. Washington, D.C.: U.S. Govt Printing Office.

Courtney, R. C. & White, R. S. 1986 Anomalous heat flow and geoid across the Cape Verde Rise: evidence for dynamic support from a thermal plume in the mantle. *Geophys. Jl R. astr. Soc.* **87**, 815–867 and microfiche GJ 87/1.

Cox, K. G. 1980 A model for flood basalt volcanism. *J. Petrol.* **21**, 629–650.

Crough, S. T. 1981 Free-air gravity over the Hoggar Massif, north-western Africa: evidence for alteration of the lithosphere. *Tectonophys.* **77**, 189–202.

Dash, B. P., Ball, M. M., King, G. A., Butler, L. W. & Rona, P. A. 1976 Geophysical investigation of the Cape Verde archipelago. *J. geophys. Res.* **81**, 5249–5259.

Davis, E. E. 1982 Evidence for extensive basalt flows on the sea floor. *Geol. Soc. Am. Bull.* **93**, 1023–1029.

Detrick, R. S. & Watts, A. B. 1979 An analysis of isostasy in the world's oceans: three aseismic ridges. *J. geophys. Res.* **84**, 3637–3655.

Duncan, R. A. & Hargraves, R. B. 1990 $^{40}Ar/^{39}Ar$ geochronology of basement rocks from the Mascarene Plateau, the Chagos Bank, and the Maldives Ridge. In *Proc. ODP, Sci. Results* (ed. R. A. Duncan *et al.*), vol. 115, pp. 43–52. College Station, Tx (Ocean Drilling Program).

Ebinger, C. J., Bechtel, T. D., Forsyth, D. W. & Bowin, C. O. 1989 Effective elastic plate thickness beneath the East African and Afar Plateaus and dynamic compensation of the uplifts. *J. geophys. Res.* **94**, 2883–2901.

Ellam, R. M. 1992 Lithospheric thickness as a control on basalt geochemistry. *Geology* **20**, 153–156.

Francis, T. J. G. & Shor, G. G. 1966 Seismic refraction measurements in the northwest Indian Ocean. *J. geophys. Res.* **71**, 427–449.

Gebrande, H., Miller, H. & Einarsson, P. 1980 Seismic structure of Iceland along RRISP-Profile I. *J. Geophys.* **47**, 239–249.

Green, W. V., Achauer, U. & Meyer, R. P. 1991 A three-dimensional seismic image of the crust and upper mantle beneath the Kenya Rift. *Nature, Lond.* **354**, 199–203.

Gripp, A. E. & Gordon, R. G. 1990 Current plate models relative to the hotspots incorporating the NUVEL-1 global plate motion model. *Geophys. Res. Lett.* **17**, 1109–1112.

Helmberger, D. V. & Morris, G. B. 1970 A travel time and amplitude interpretation of a marine refraction profile: transformed shear waves. *Bull. seis. Soc. Am.* **60**, 593–600.

Holik, J. S. & Rabinowitz, P. D. 1992 Effects of Canary hotspot volcanism on structure of oceanic crust off Morocco. *J. geophys. Res.* **96**, 12093–12067.

Houseman, G. A. 1990 The thermal structure of mantle plumes: axisymmetric or triple-junction? *J. Geophys. Int.* **102**, 15–24.

Karson, J. A. & Curtis, P. C. 1989 Tectonic and magmatic processes in the Eastern Branch of the East African Rift and implications for magmatically active continental rifts. *J. African Earth Sci.* **8**, 431–453.

KRISP Working Group 1991 Structure of the Kenya Rift from seismic refraction. *Nature, Lond.* **325**, 239–242.

LeCheminant, A. N. & Heaman, L. M. 1989 Mackenzie igneous events, Canada: Middle Proterozoic hotspot magmatism associated with ocean opening. *Earth planet. Sci. Lett.* **96**, 38–48.

Lindwall, D. A. 1988 A two-dimensional seismic investigation of crustal structure under the Hawaiian Islands near Oahu and Kauai. *J. geophys. Res.* **93**, 12107–12122.

Lister, J. R. 1990 Buoyancy-driven fluid fracture: similarity solutions for the horizontal and vertical propagation of fluid-filled cracks. *Fluid Mech.* **217**, 213–239.

Lister, J. R. & Kerr, R. C. 1991 Fluid-mechanical models of crack propagation and their application to magma transport in dykes. *J. geophys. Res.* **96**, 10049–10077.

Mahoney, J., Nicollet, C. & Dupuy, C. 1991 Madagascar basalts: tracking oceanic and continental sources. *Earth planet. Sci. Lett.* **104**, 350–363.

Martin, B. S. 1989 The Roza Member, Columbia River Basalt Group: Chemical stratigraphy and flow distribution. In *Volcanism and tectonism in the Columbia River flow-basalt province* (ed. S. P. Reidel & P. R. Hooper), vol. 239, pp. 85–104. Geological Society of America Special Paper.

Morgan, J. V., Barton, P. J. & White, R. S. 1989 The Hatton Bank continental margin. III. Structure from wide-angle OBS and multichannel seismic refraction profiles. *Geophys. J. Int.* **98**, 367–384.

McKenzie, D. P. 1984 A possible mechanism for epeirogenic uplift. *Nature, Lond.* **307**, 616–618.

McKenzie, D. P. & Bickle, M. J. 1988 The volume and composition of melt generated by extension of the lithosphere. *J. Petrol.* **29**, 625–679.

McKenzie, D. & O'Nions, R. K. 1991 Partial melt distributions from inversion of rare earth element concentrations. *J. Petrol.* **32**, 1021–1091.

Oxburgh, E. R. & Parmentier, E. M. 1977 Compositional and density stratification in oceanic lithosphere – causes and consequences. *J. geol. Soc. Lond.* **133**, 343–355.

Parsons, B. & Richter, F. M. 1981 Mantle convection and the oceanic lithosphere. In *The sea*, vol. 7, *The oceanic lithosphere* (ed. C. Emiliani), pp. 73–117. New York: Wiley.

Recq, M., Brefort, D., Malod, J. & Veinante, J.-L. 1990 The Kerguelen Isles (southern Indian Ocean): new results on deep structure from refraction profiles. *Tectonophys.* **182**, 227–248.

Ritzert, M. & Jacoby, W. R. 1985 On the lithospheric seismic structure of Reykjanes Ridge at 62.5° N. *J. geophys. Res.* **90**, 10117–10128.

Royer, J.-V., Sclater, J. G. & Sandwell, D. T. 1989 A preliminary tectonic fabric chart of the Indian Ocean. *Proc. Indian Acad. Sci. (Earth planet. Sci.)* **98**, 7–24.

Saunders, A. D. 1985 Geochemistry of basalts from the Nauru basin, Deep Sea Drilling Project legs 61 and 89: Implications for the origin of oceanic flood basalts. In *Initial Reports DSDP 89* (ed. R. Moberley *et al.*), pp. 499–517. Washington, D.C.: U.S. Govt Printing Office.

Schlich, R. 1982 The Indian Ocean: Aseismic ridges, spreading centres, and ocean basin. In *The ocean basins and margins*, vol. 6, *The Indian Ocean* (ed. A. E. M. Nairn & F. G. Stehli), pp. 51–147. New York: Plenum Press.

Schmincke, H.-U. 1982 Volcanic and chemical evolution of the Canary Islands. In *Geology of the northwest African Continental Margin* (ed. U. von Rad, K. Hinz, M. Sarnthein & E. Seibold), pp. 273–306. New York: Springer-Verlag.

Schubert, G. & Sandwell, D. 1989 Crustal volumes of the continents and of oceanic and continental submarine plateaus. *Earth planet Sci. Lett.* **92**, 234–246.

Sinha, M. C., Louden, K. E. & Parsons, B. 1981 The crustal structure of the Madagascar Ridge. *Geophys. Jl R. astr. Soc.* **66**, 351–377.

Stolper, E. & Walker, D. 1980 Melt density and the average composition of basalt. *Contributions Mineralogy Petrology* **74**, 7–12.

Tarduno, J. A., Sliter, W. V., Kroenke, L., Leckie, M., Mayer, H., Mahoney, J. J., Musgrave, R., Storey, M. & Winterer, E. L. 1991 Rapid formation of Ontong Java Plateau by Aptian mantle plume volcanism. *Science, Wash.* **254**, 399–403.

ten Brink, U. S. & Brocher, T. M. 1987 Multichannel seismic evidence for a subcrustal intrusive complex under Oahu and a model for Hawaiian volcanism. *J. geophys. Res.* **92**, 13687–13707.

Tolan, T. L., Reidel, S. P., Beeson, M., Anderson, J. L., Fecht, K. R. & Swanson, D. A. 1989 Revisions to the estimates of the aerial extent and volume of the Columbia River Basalt Group. In *Volcanism and tectonism in the Columbia River flood-basalt province* (ed. S. P. Reidel & P. R. Hooper), pp. 1–20, Geol. Soc. Amer. Spec. Pap. 239. Boulder, Colorado.

Watson, S. & McKenzie, D. 1991 Melt generation by plumes: a study of Hawaiian volcanism. *J. Petrol.* **32**, 501–537.

Watts, A. B. & ten Brink, U. S. 1989 Crustal structure, flexure and subsidence history of the Hawaiian Islands. *J. geophys. Res.* **94**, 10473–10500.

Watts, A. B., McKenzie, D. P., Parsons, B. & Roufosse, M. 1985*a* The relationship between gravity and bathymetry in the Pacific Ocean. *Geophys. Jl R. astr. Soc.* **83**, 263–298.

Watts, A. B., ten Brink, U. S., Buhl, P. & Brocher, T. M. 1985*b* A multichannel seismic study of lithospheric structure across the Hawaiian-Emperor seamount chain. *Nature, Lond.* **315**, 105–111.

White, R. R. 1992 Crustal structure and magmatism of North Atlantic continental margins. *J. Geol. Soc.* **149**, 841–854.

White, R. S. & McKenzie, D. P. 1989 Magmatism at rift zones: the generation of volcanic continental margins and flood basalts. *J. geophys. Res.* **94**, 7685–7729.

White, R. S., McKenzie, D. & O'Nions, R. K. 1992 Oceanic crustal thickness from seismic measurements and rare earth element inversions. *J. geophys. Res.* (In the press.)

Williams, L. A. J. 1972 The Kenyan Rift volcanics: a note on volumes and chemical composition. *Tectonophys.* **15**, 83–96.

Zucca, J. J., Hill, D. P. & Kovach, R. L. 1982 Crustal structure of Mauna Loa volcano, Hawaii, from seismic refraction and gravity data. *Bull. seis. Soc. Am.* **72**, 1535–1550.

Continental magmatic underplating

By K. G. Cox

Department of Earth Sciences, Parks Road, Oxford OX1 3PR, U.K.

Three arguments based on geological evidence are put forward to support the importance of magmatic underplating processes during continental flood basalt vulcanism. (1) Petrological evidence of gabbro fractionation in erupted basaltic sequences allows estimates to be made of the minimum total mass of concealed cumulate material, which is retained in deep crustal magma chambers, possibly along the Moho, and is comparable in amount to the erupted material. (2) In the Karoo province (southern Africa) large volumes of rhyolite along the S.E. continental margin were generated from basaltic precursors, either as partial melts of already-emplaced solid basic material or as crystal fractionation products of large volumes of basic magma. In either case very substantial volumes of concealed basic rocks are at least locally implied. (3) Studies of geomorphology suggest that the area of the Karoo province experienced at least 1 km of permanent uplift associated with the vulcanism. This appears to be the consequence of the emplacement of an underplated gabbroic layer *ca.* 5 km thick.

1. Introduction

Because of the relatively low densities of crustal rocks, basaltic magmas generated beneath continental areas are probably frequently trapped at or near the Moho, or within the crust, or in complex crust–mantle transition zones. This is the phenomenon that has come to be known as 'underplating'. Thus, although large volumes of magma reach the surface, for example in continental flood basalt (CFB) provinces, a proportion, in some cases possibly a large proportion, solidifies at depth, and the products remain hidden (Cox 1980). The magmas of principal interest are in almost all cases the tholeiites generated in large quantities during CFB vulcanism. Any mantle-derived magma can give rise to underplate material, but the CFB tholeiites are overwhelmingly the most significant in terms of the volumes produced. It is important to understand underplating for two reasons. First, a full appreciation of the amounts of melt generated in an igneous event depends on the quantification of the phenomenon, and second, underplating can lead to significant crustal growth and thickening, which would be unsuspected merely from the examination of the surface products.

There are many ways of approaching the underplating problem, including the theoretical study of the physical problems concerned, the seismological investigation of the lower crust and uppermost mantle, and the study of xenoliths of deep crustal or shallow upper mantle origin (O'Reilly 1988). The present paper, however, reviews the evidence that can be derived from the petrological and geochemical investigation of erupted volcanic sequences in CFB provinces, and also makes a preliminary assessment of the geomorphological evidence for underplating in such areas.

Phil. Trans. R. Soc. Lond. A (1993) **342**, 155–166

Printed in Great Britain

Under the petrology/geochemistry heading, two lines of investigation are pursued. The first may be termed the problem of the missing cumulates (erupted lavas show conclusive evidence of having undergone significant amounts of fractional crystallization at crustal levels. Where are the crystals now? And how much of this material is there?). The second concerns the generation of rhyolites on continental margins. In some cases (e.g. S.E. Africa during the early Jurassic) huge volumes of rhyolite were either generated from basaltic source rocks by partial melting, or were products of the fractional crystallization of large volumes of basaltic magma. In either case, very large volumes of hidden basaltic material are implied.

Geomorphology provides the third line of argument. Long-lived uplift of within-plate continental areas can in some cases be firmly linked to CFB vulcanism, and is difficult to explain without recourse to crustal thickening consequent upon unseen magmatic additions (McKenzie 1984).

2. The missing cumulates

Over the years there has been vigorous argument about the major element composition of primary magmas from the mantle in continental areas, especially about exactly how picritic (olivine-rich or MgO-rich) they might be. Picritic basalts are locally important in some CFB provinces such as the North Atlantic Tertiary province (Clarke 1970), the Karoo province (Bristow 1984), and the Deccan (Krishnamurthy & Cox 1977), and have been assigned a parental role in the generation of the more-evolved, and much more wide-spread, low-MgO basalts typical of these areas (O'Hara 1965; Cox 1980). Conversely, some have argued that the evolved basalts are themselves primary (Wilkinson & Binns 1977; Wright *et al.* 1989). Fortunately, for most of the purposes of the present paper, the precise Mg-content of the primary magmas is unimportant. Picritic magmas readily evolve to basaltic compositions by the fractionation of ferromagnesian minerals, principally olivine. The cumulates produced are not unimportant, but nevertheless they probably rarely constitute more than 25 wt % of the original mass of magma generated (see Cox (1980) for calculation of the amount of olivine fractionation required to generate a basaltic liquid from a typical picritic liquid). In principle, if it were possible to estimate the total amount of *basaltic* material present in a province (i.e. surface eruptives, high-level intrusives, gabbroic underplate material) it would be necessary to augment the total by this factor. However, in terms of crustal growth and thickening, cumulates consisting only of ferromagnesian minerals qualify as mantle, not crust. From the point of view of this question, the important issue is thus, how much plagioclase-bearing cumulate material is present? Such material will count as an addition to the crust, because plagioclase is the low-density phase that most-distinguishes the crust from the mantle.

Most continental tholeiitic magmas begin to crystallize the gabbroic assemblage olivine + clinopyroxene + plagioclase when the MgO content of the liquid has been reduced to *ca.* 7 wt % by the fractionation of ferromagnesian phases. The proportions in which the gabbroic phases crystallize is *ca.* 10–15 % olivine, 30–40 % clinopyroxene, and 50–55 % plagioclase (see Cox & Mitchell 1988; Harris *et al.* 1990). There are exceptions, but they are rare. For example, an abnormally high potassium content (*ca.* 2 wt %) suppresses plagioclase crystallization, and in such cases the MgO content of the liquid can be reduced to as little as 4 wt % without the appearance of this phase (Cox & Bristow 1984).

Phil. Trans. R. Soc. Lond. A (1993)

Many studies give evidence of gabbro fractionation in CFB sequences, and highly distinctive inter-element relationships are seen. Most characteristic is absolute iron-enrichment with falling MgO, because the Fe-free phase, plagioclase, constitutes more than half of the fractionating assemblage. Other typical features are a strong positive correlation between MgO and CaO, and a gentle decline of Al_2O_3 with falling MgO. Amongst the trace elements the behaviour of Sr is particularly distinctive, in that it shows little variation, as a consequence of a bulk K_D close to unity in the fractionating assemblage.

One of the most important features of the fractional crystallization of a gabbroic assemblage from a basaltic liquid is that the removal of relatively large amounts of crystalline material is accompanied by rather small changes in the concentrations of many elements in the residual liquid. For these elements, the bulk composition of the cumulate material is fairly close to that of the liquid. Hence, lava sequences which at first sight look compositionally relatively uniform, may in fact imply the existence of significant quantities of associated cumulates.

To determine the minimum amount of plagioclase-bearing cumulus material associated with an eruptive sequence it is necessary to estimate the average composition of the whole sequence, the composition of the parent magma when it first begins to crystallize the gabbroic assemblage, and the bulk composition of the cumulate assemblage itself. Mass balance requirements for any individual element then lead to the relationship:

$$C_P = C_L X_L + C_C X_C,$$

where C_P, C_L and C_C are the concentrations of the element in parent magma, erupted liquid, and cumulate respectively, and X_L and X_C are the mass fractions of the erupted liquid and the cumulates, totalling the original unit mass of parental liquid. In practice the total mass of the erupted sequence is rarely determinable with any degree of accuracy, because of unknown amounts lost by erosion or concealed by younger deposits. Hence it is not possible to determine the absolute cumulate mass. However, it is extremely useful to determine the relative masses of the erupted sequence and the cumulates, i.e. the ratio X_C/X_L, which is derived from the familiar 'lever rule' equations:

$$X_L = (C_P - C_C)/(C_L - C_C),$$
$$X_C = (C_L - C_P)/(C_L - C_C).$$

It is best to carry out the calculations by using the concentrations of major elements rather than incompatible trace- and minor-elements, which often appear to show anomalous degrees of enrichment during fractionation. One probable reason for this is that the parental magma, as it begins to fractionate gabbro, may itself already be variable in its incompatible element content (see Gill *et al.* 1988), which is equivalent to having an extra enrichment-factor built in, and has the effect of magnifying the apparent enrichment during later fractional crystallization.

The major element calculation can be approached via generalized numerical modelling (Cox 1980) but the example given below is based on one of the much-improved data-sets now available. The rocks considered are those of the Ambenali Formation in the Deccan Traps, selected because they are free from significant crustal contamination, and characterized by numerous analyses. Devey (1986) for example systematically sampled road sections in the Western Ghats, analysing more than 100 samples from the formation. Using average phenocryst compositions he modelled gabbro fractionation for the Ambenali data-set, concluding that the range of compositions erupted (excluding a few more basic rocks which were obviously not on the gabbro fractionation trend) could be generated by the fractionation of olivine, clinopyroxene, and plagioclase in the proportions $14:35:51$ from a parent magma containing 7.58 wt % MgO. The most-evolved erupted composition contains 5.64 wt % MgO, and is generated by the crystallization of about 42.5 wt % of the original liquid. In the present study I have selected one of Devey's road sections, at Devrukh, because this is the only one that includes the complete Ambenali section, here 480 m thick. The section is represented by 22 analyses. The average composition of the section can be calculated by weighting each analysis according to the vertical sample spacing, assuming that it represents the column of rock half way up to the overlying sample site, and half way down to the underlying site. The average MgO content calculated in this way is 6.21 wt %. This is of course more basic than Devey's most-evolved composition, so the amount of fractionation required is less than the 42.5 % crystallization required above. In fact it works out at ca. 33 %.

Hence, $X_C/X_L \approx 0.5$, which is of course subject to slight errors because of uncertainties in the composition of both the parent and the cumulates. Nevertheless, a significant quantity of hidden cumulus material is implied, and taking the ratio at face value the implication is that for every cubic kilometre of Ambenali magma erupted, another 0.5 km³ of cumulates exists. In this case the formation is ca. 480 m thick in the area sampled, implying the equivalent of a 240 m thick layer of hidden cumulates.

Calculations of this sort do not characteristically yield very large values for the ratio X_C/X_L, though Cox (1980) estimated that many erupted CFB sequences were probably accompanied by approximately equivalent amounts of hidden cumulates. Estimates of the latter become larger as the rocks become more evolved, and there are for example substantial tracts of CFB provinces made up of more evolved lavas than the Ambenali Formation (e.g. an average MgO of ca. 5.5 wt % for the Sabie River Basalt Formation of the Zululand is reported by Duncan et al. (1984)). However, the significance of these estimates does not lie in the fact that the amounts of hidden cumulate indicated are particularly large. Rather, they demonstrate the existence of hidden magma chambers of significant size, that the conditions necessary for underplating existed, and that at least some underplating must have taken place. The calculations furthermore can only give minimum estimates of the amount of hidden material. They depend on the assumption that all fractionated liquids produced are available for sampling on the surface. Unknown additional amounts of magma may have solidified completely in the sub-surface environment, leaving no trace in the eruptive record.

In the next section a different line of petrological evidence will be examined, suggesting that, at least locally on continental margins, massive amounts of underplate material may exist, probably in much larger volumes than indicated by the calculations above.

3. Rhyolites on continental margins: evidence from S.E. Africa

Several instances are known of CFB provinces that contain substantial amounts of acid rocks, often in the form of relatively small plutons, but occasionally in the form of rhyolitic extrusives. At the present level of erosion the latter are developed on a small scale in the Deccan (e.g. on Salsette Island near Bombay (Sethna & Battiwala 1980)), on an areally extensive scale, though not very thick, in the Paraná Province

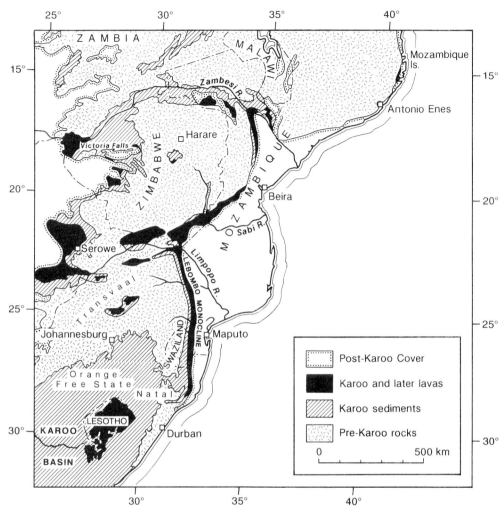

Figure 1. Geological map of South-East Africa showing distribution of Karoo rocks.

(Bellieni *et al.* 1984), and on a truly huge scale in the southeastern part of the Karoo Province in southern Africa (Cleverly 1979; Cleverly *et al.* 1984)). In the latter area the rhyolites are exposed continuously over a distance of *ca.* 600 km along the Lebombo monocline (see figure 1) and a maximum thickness of as much as 7.5 km has been estimated in Swaziland from surface dip measurements and outcrop width (Cleverly 1979). Had the monocline been below sea-level, it would certainly have been identified as a sequence of seaward dipping reflectors by marine geophysicists. For whatever reasons (see Cox 1992) it is exposed above sea-level now, and is available for inspection. A geological map of the Lebombo in Swaziland and neighbouring parts of Mozambique is given in figure 2 and illustrates the fundamental features of the present argument. Using observed dip measurements on the surface, the volcanic sequence exposed appears to be approximately 15 km thick (see figure 3), of which about half consists of rhyolite. The volcanic sequence rests in the west on Karoo sediments overlying Archaean basement. To the east the nature of the basement is unknown, though basement is faulted up on the seaward side of the

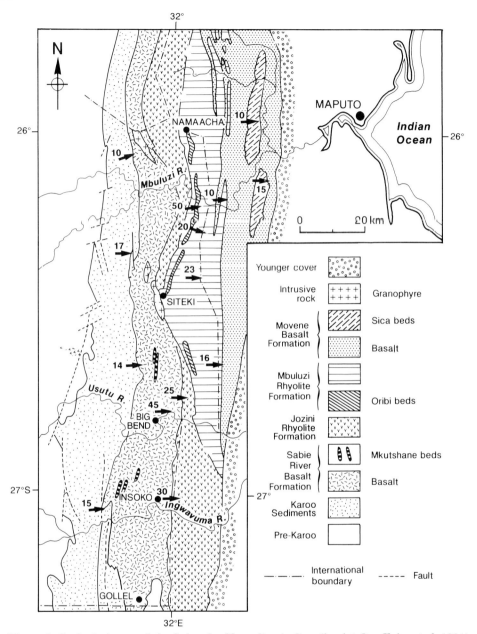

Figure 2. Geological map of the Lebombo Monocline in Swaziland (after Eales *et al.* 1984).

monocline in Zululand at the southern end of the province, forming a half-graben. The area east of the monocline has been referred to as the Mozambique Thinned Zone (Cox 1992), an area possibly of new oceanic crust, or of thinned continental crust, or a mixture of both, overlain by younger sediments and produced during the early stages of the break-up of Gondwanaland in the L. Jurassic. The Lebombo can perhaps not strictly be regarded as a continental margin, but it is clearly the boundary between crustal areas respectively unaffected by and strongly affected by the early stages of continental break-up. It is not possible to estimate the total

W   E

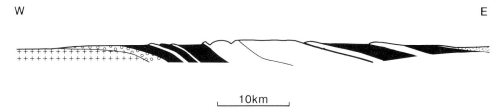

Figure 3. Sketch section across the Lebombo along the line of the Mbuluzi R. (see figure 2). From west to east ornaments are: crosses, Archaen basement; circles, Karoo sediments; black, basalts; un-ornamented, rhyolites; stipple, post-Karoo cover. Horizontal and vertical scales are the same.

volume of the rhyolites because extents are unknown both up-dip (to the west) and down-dip. They are, however, absent from the inland Karoo volcanic sequences in the Transvaal and Lesotho, approximately 200 km west of the monocline. On the seaward side the extent to which half-graben systems, and general thinning towards the ocean, may affect the sequence is unknown.

The origin of the Lebombo rhyolites has been extensively discussed in the literature, since Manton (1968) carried out the first Sr-isotopic determinations and discovered that they display initial ratios typical of mantle-derived rocks ($^{87}Sr/^{86}Sr_I = 0.7044 \pm 2$ (Bristow *et al.* 1984)) and indeed ratios similar to the basalts which they overlie. A small sequence of stratigraphically lower rhyolites (the Mkutshane Beds, see figure 2), are in complete geochemical contrast to the main rhyolite sequence, and have all the expected geochemical features of remelts derived from Archaean basement. The existence of the Mkutshane Beds thus reinforces the conviction that the main sequence is, somehow, related to its associated basalts rather than to remelting of the basement.

Rhyolites cannot be derived from the mantle in a one-stage process, but they can be generated by the further processing of first-stage products, e.g. by the remelting of basalt or by the fractional crystallization of basaltic magma. Betton (1978; and see Cleverly *et al.* 1984) carried out detailed modelling calculations and concluded that most of the rhyolites of the main sequence were generated by the partial melting of source rocks similar to Karoo dolerites, having a mineralogy dominated by clinopyroxene and plagioclase, with subordinate magnetite, quartz, and potash feldspar. For various specific rhyolite types the amount of melting required varied from 11 to 16 wt % of the source rock. Essentially similar conclusions have been reached for the Deccan rhyolites of Salsette (Lightfoot *et al.* 1987). Betton also modelled fractional crystallization of typical Swaziland basaltic magmas and concluded that this too was a permissible mechanism for generating the rhyolites. It was however regarded as a less-likely origin, because the rhyolites occur only in the specific setting of the Lebombo – implying some sort of special conditions of formation – not in the continental interior, where basalts alone occur. If fractional crystallization were the answer, then at least some rhyolite should be expected in all areas.

However, from the point of view of the present argument, it is actually immaterial whether the rhyolites were generated by the remelting of previously solidified basaltic material or by the fractional crystallization of basaltic magma. Nor is the mechanism by which remelting might occur particularly relevant (Betton (1978), for example, suggested that uplift, implying decompression, of hot underplated material during crustal thinning might suffice. Alternatively, injection of fresh magma into the underplate zone might promote partial melting). In either case, the rhyolites erupted onto the surface require the existence of perhaps 6–10 times as much basaltic material, either as their source rocks or as their accompanying gabbroic cumulates. A rhyolite sequence of this type, up to 7.5 km thick, implies the existence of very large volumes of such hidden material, which was probably emplaced as an underplate, analogous to the prisms of high-velocity lower continental crust identified beneath some volcanic continental margins. For example the prism identified by White *et al.* (1987) on the Hatton Bank margin in the N. Atlantic, and interpreted by them as underplated basaltic material, extends for *ca.* 100 km normal to the margin, and reaches a thickness of 15 km. Partial remelting of a source of this size would clearly be capable of generating impressive quantities of rhyolite.

To reinforce the argument, in the next section the topographic evidence for uplift in southeast Africa will be interpreted to imply that in fact underplated material probably extends for several hundred kilometres into the continental interior beneath pre-existing crust.

4. The topographic argument

Uplift of within-plate continental areas can be generated by the dynamic and thermal effects of plume activity, or by crustal thickening, which in the absence of folding and thrusting is perhaps most likely to take place by the magmatic addition of rocks with densities lower than that of the mantle (McKenzie 1984). Dynamic plume effects can, however, be discounted in most CFB provinces, because hot spot tracks indicate that plumes, even if still active, are now large distances away from the provinces created (e.g. the Deccan plume now under Réunion, the Paraná-Etendeka plume now under Tristan da Cunha). Equally, thermal effects are unlikely still to be in evidence in Mesozoic provinces such as the Karoo and the Paraná, because the time constants of thermal decay are too short. Thermal affects are probably, however, still of some significance in the early Tertiary N. Atlantic province (White & McKenzie 1989), and may still have some small influence in the Deccan.

The Karoo province, however, provides evidence that the uplift which took place either during or after vulcanism is still clearly expressed, i.e. it is the type referred to as 'permanent uplift' (McKenzie 1984; in contrast to 'impermanent' uplifts caused by dynamic and thermal effects), and seems therefore most likely to be related to magmatic crustal thickening. The uplift involved is the 'surface uplift' of Molnar & England (1990), i.e. the average surface elevation was increased, and it is not to be confused with the upward movement of bed-rock during the isostatic response to erosion, during which average surface elevation decreases. The argument for southeast Africa has been discussed by Cox (1989). Briefly, the salient points are:

1. The area is topographically very high compared with most anorogenic continental regions (maximum elevation *ca.* 3.5 km, large areas at 1.5–2 km above sea level).

2. Karoo sedimentation immediately before the onset of vulcanism was largely aeolian and fluviatile, but it was unlikely to have taken place at more than *ca.* 0.5 km above sea-level.

3. Over large areas of the Transvaal and Orange Free State the Karoo lavas have largely been eroded away but the underlying sedimentary sequences are preserved. Thus any contribution to topographic height made by the lava pile itself has been

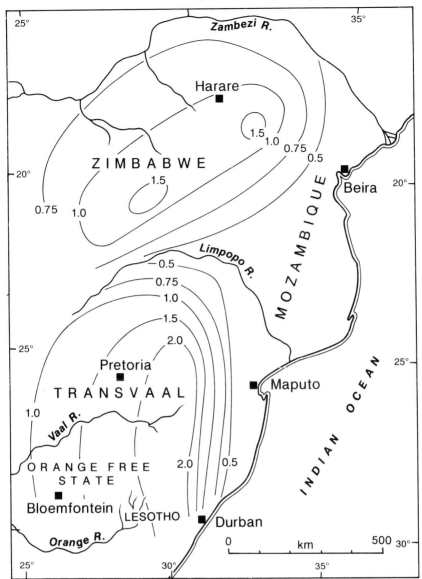

Figure 4. Sketch map showing estimated elevation above sea-level of the present-day sub-Karoo Basalt unconformity in S.E. Africa. Reference to figure 1 will indicate degree of certainty (i.e. where the unconformity is actually exposed). In some cases high-points have been inserted in the map so that the unconformity can pass above high areas of pre-Karoo rocks, e.g. 200 km S.E. of Harare (Marandelas), 400 km S.W. of Harare (Matopo Hills), and 200 km N.E. of Pretoria (the Berg). Elevations above 0.5 km above sea-level are probably the consequence of surface uplift. The half-dome pattern of the uplift (see Cox 1989) is bisected by the Limpopo rift.

removed. In the absence of any other factor, isostatic readjustment should have restored the sub-basalt unconformity to its original elevation. It is, however, now preserved patchily, or can be inferred to have been recently only a little above the present ground surface, over wide areas of South Africa and Zimbabwe at more than 1 km, and in some areas, more than 2 km above sea-level (see figure 4). Evidently there *is* another factor, and that is likely to be the contribution to crustal thickness made by the underplate. For example, in the area of figure 4 where the unconformity is inferred to be 2 km above sea-level, at least 1 km of uplift can safely be invoked. The density assumptions of McKenzie (1984; i.e. density of *gabbroic* underplated material of 2.7 Mg m^{-3}, mantle density of 3.3 Mg m^{-3}), imply an underplated layer including *ca.* 5.5 km of gabbro in this area. The total amount of underplated material is of course likely to be substantially higher than this because of additional ultramafic cumulates. White & McKenzie (1989), for example, assumed an overall density of 3.0 Mg m^{-3} for the underplated material, which leads to an increase by a factor of two in the calculated thickness. However, as pointed out earlier, the included ultramafic material does not affect the uplift argument.

In the area discussed above, the maximum preserved thickness of surface lavas is *ca.* 1.5 km, e.g. in Lesotho (Cox & Hornung 1967), suggesting that the underplate is much thicker than the surface sequence. Furthermore, the area affected by uplift appears to be much wider than those including typical uplifted rift-shoulders (e.g. East Africa) for which there might be other explanations. There is a strong implication that at least *some* underplated material was emplaced far inland of the continental margin.

5. General remarks

This inference above conjours up interesting speculations about crustal growth over geological time spans. During the past 250 Ma, there have been half a dozen or more CFB episodes akin to the Karoo (e.g. Siberia, Paraná, Deccan, N. Atlantic, Ethiopia–Yemen, Columbia River). On that basis, if such a mechanism operated in earlier times, a not insignificant fraction of the continental crust could perhaps have been generated by underplating in addition to the marginal accretion of island arcs. White & McKenzie (1989) estimated that the crustal accretion rate from extensional magmatism was *ca.* 0.4 km^3 a^{-1}, compared with an average Phanerozoic rate of perhaps 1 km^3 a^{-1}. However, for these figures to be meaningful, the high density lower crust produced by extensional magmatism must not be preferentially lost back into the mantle.

Since many of the arguments presented have been based on the Karoo province the extent to which this case acts as a general model for CFB vulcanism is of some interest. In general, it has to be said, the effects appear to be smaller in most other provinces, because the areas simply are not so topographically high as S.E. Africa. Areas comparable in elevation do exist in Ethiopia and Yemen, where extensive tracts are more than 2 km above sea-level. However, these are likely to be still dynamically supported by the Afar plume. An additional general problem lies in the fact that quantification of surface uplift is extremely difficult, because it is necessary to estimate both the original elevation of the surface onto which the basalts were erupted, and the present elevation at which the sub-basalt unconformity would lie if the basalts had been just eroded off, and the underlying topography remained as an un-canyoned plateau surface. The latter is a requirement because any substantial dissection of the sub-basalt rocks results in the renewed uplift of the unconformity, e.g. preserved locally on interfluves, by isostatic adjustment.

Surface uplift has, however, almost certainly affected areas of the North Atlantic Tertiary Province on both sides of the present ocean, in Greenland, Norway, and Scotland, but it is not clear at present how much of it is 'permanent' as in the Karoo case. In Scotland, Watson (1985) put forward a closely argued case, demonstrating that the western part of the Scottish craton was uplifted in the Eocene by

0.5–1.5 km, an event which she specifically related to the co-eval N. Atlantic vulcanism. Brooks (1979, 1985) has described the Eocene uplift of the sub-basalt unconformity inland of Skaergaard in E. Greenland, which resulted in an extensive inlier of basement gneisses. However, a second uplift-episode later in the Tertiary also seems to have taken place, suggesting that rejuvenation of initial uplifts is possible. In Norway, Torske (1975) has ascribed reversals of drainage directions to plume activity associated with the opening of the N. Atlantic in the Mesozoic, i.e. *before* the North Atlantic vulcanism, but there is also an episode of Tertiary uplift. The latter is, however, difficult to quantify by using the arguments developed above. Norway is now devoid of the volcanic sequence, if it were formerly present, and is deeply dissected, probably making the isostatic effects important. Nevertheless, it is clear that the geomorphology has a continuing role to play in the study of the vulcanism of within-plate continental regions.

References

Bellieni, G., Brotzu, P., Comin-Chiaramonti, P., Ernesto, M., Melfi, A. J., Pacca, I. G. & Piccirillo, E. M. 1984 Flood basalt to rhyolite suites in the southern Paraná plateau (Brazil): palaeomagnetism, petrogenesis and geodynamic implications. *J. Petrol.* **25**, 579–618.

Betton, P. J. 1978 Geochemistry of Karoo volcanic rocks in Swaziland. D.Phil. thesis, Oxford University.

Bristow, J. W. 1984 Picritic rocks of the North Lebombo and south-east Zimbabwe. In *Petrogenesis of the volcanic rocks of the Karoo province* (ed. A. J. Erlank), *Spec. Publ. geol. Soc. S. Afr.* **13**, 105–123.

Brooks, C. K. 1979 Geomorphological observations at Kangerdlugssuaq, East Greenland. *Greenland Geosci.* **1**, 3–21.

Brooks, C. K. 1985 Vertical crustal movements in the Tertiary of central East Greenland: a continental margin hot-spot. *Z. Geomorph. N.F., Suppl.-Bd.* **54**, 101–117.

Clarke, D. B. 1970 Tertiary basalts from Baffin Bay; possible primary magma from the mantle. *Contr. Miner. Petrol.* **25**, 203–204.

Cleverly, R. W. 1979 The volcanic geology of the Lebombo monocline in Swaziland. *Trans. geol. Soc. S. Afr.* **82**, 343–348.

Cleverly, R. W., Betton, P. J. & Bristow, J. W. 1984 Geochemistry and petrogenesis of the Lebombo rhyolites. In *Petrogenesis of the volcanic rocks of the Karoo province* (ed. A. J. Erlank), *Spec. Publ. geol. Soc. S. Afr.* **13**, 171–194.

Cox, K. G. 1980 A model for flood basalt vulcanism. *J. Petrol.* **21**, 629–650.

Cox, K. G. 1989 The role of mantle plumes in the development of continental drainage patterns. *Nature, Lond.* **342**, 873–877.

Cox, K. G. 1992 Karoo igneous activity, and the early stages of the break-up of Gondwanaland. In *Magmatism and the causes of continental break-up* (ed. B. Storey & A. Alabaster), *Spec. Publ. geol. Soc. Lond.*

Cox, K. G. & Bristow, J. W. 1984 The Sabie River Basalt formation of the Lebombo Monocline and south-east Zimbabwe. In *Petrogenesis of the volcanic rocks of the Karoo province* (ed. A. J. Erlank), *Spec. Publ. geol. Soc. S. Afr.* **13**, 125–147.

Cox, K. G. & Mitchell, C. 1988 Importance of crystal settling in the differentiation of Deccan Trap basic magmas. *Nature, Lond.* **333**, 447–449.

Cox, K. G. & Hornung, G. 1966 The petrology of the Karoo basalts of Basutoland. *Am. Mineralogist* **51**, 1414–1432.

Devey, C. W. 1986 Stratigraphy and geochemistry of the Deccan Trap lavas, western India. D.Phil. thesis, Oxford University.

Duncan, A. R., Erlank, A. J. & Marsh, J. S. 1984 Regional geochemistry of the Karoo igneous province. In *Petrogenesis of the volcanic rocks of the Karoo province* (ed. A. J. Erlank), *Spec. Publ. geol. Soc. S. Afr.* **13**, 355–388.

Gill, R. C. O., Neilsen, T. F. D., Brooks, C. K. & Ingram, G. A. 1988 Tertiary volcanism in the Kangerdlugssuaq region, E. Greenland: trace element geochemistry of the Lower Basalts and tholeiitic dyke swarms. In *Early Tertiary vulcanism and the opening of the NE Atlantic* (ed. A. C. Morton & L. M. Parson), *Geol. Soc. Spec. Publ.* **39**, 161–179.

Harris, C., Marsh, J. S., Duncan, R. A. & Erlank, A. J. 1990 The petrogenesis of the Kirwan basalts of Dronning Maud Land, Antarctica. *J. Petrol.* **31**, 341–369.

Krishnamurthy, P. & Cox, K. G. 1977 Picrite basalts and related lavas from the Deccan Traps of western India. *Contr. Miner. Petrol.* **62**, 53–75.

Lightfoot, P. C., Hawkesworth, C. J. & Sethna, F. F. 1987 Petrogenesis of rhyolites and trachytes from the Deccan Trap: Sr, Nd, and Pb isotope and trace element evidence. *Contr. Miner. Petrol.* **95**, 44–54.

McKenzie, D. P. 1984 A possible mechanism for epeirogenic uplift. *Nature, Lond.* **307**, 616–618.

Manton, W. I. 1968 The origin of associated basic and acid rocks in the Lebombo-Nuanetsi igneous province, southern Africa, as implied by strontium isotopes. *J. Petrol.* **9**, 23–29.

Molnar, P. & England, P. 1990 Surface uplift, uplift of rocks, and exhumation of rocks. *Geology* **18**, 1173–1177.

O'Hara, M. J. 1965 Primary magmas and the origin of basalts. *Scott. J. Geol.* **1**, 19–40.

O'Reilly, S. Y. 1988 Evolution of Phanerozoic Eastern Australian lithosphere: isotopic evidence for magmatic and tectonic underplating. *J. Petrol.* (Special Lithosphere Issue), pp. 89–108.

Sethna, S. F. & Battiwala, H. K. 1980 Major element geochemistry of the intermediate and acidic rocks associated with the Deccan Trap basalts. In *Proc. 3rd Ind. geol. Cong. Poona*, pp. 281–294.

Torske, T. 1975 Possible Mesozoic mantle plume activity beneath the continental margin of Norway. *Norges geologiske undersøkelse* **322**, 73–90.

Watson, J. 1985 Northern Scotland as an Atlantic–North Sea divide. *J. geol. Soc. Lond.* **142**, 221–243.

White, R. S., Spence, G. D., Fowler, S. R., McKenzie, D. P., Westbrook, G. K. & Bowen, A. N. 1987 Magmatism at rifted continental margins. *Nature, Lond.* **330**, 439–444.

Wilkinson, J. F. G. & Binns, R. A. 1977 Relatively iron-rich lherzolite xenoliths of the Cr-diopside suite: a guide to the primary nature of anorogenic tholeiitic andesites. *Contr. Miner. Petrol.* **65**, 199–212.

Wright, T. L., Mangan, M. & Swanson, D. A. 1989 Chemical data for flows and feeder dykes of the Yakima Basalt sub-group, Columbia River Basalt Group, Washington, Oregon, and Idaho, and their bearing on a petrogenetic model. *U.S. geol. Surv. Bull.* 1821, 71 pp.

Deep structure of arc volcanoes as inferred from seismic observations

By A. Hasegawa[1], A. Yamamoto[1], D. Zhao[2], S. Hori[1] and S. Horiuchi[1]

[1] *Observation Center for Prediction of Earthquakes and Volcanic Eruptions, Faculty of Science, Tohoku University, Sendai 980, Japan*
[2] *Geophysical Institute, University of Alaska Fairbanks, Fairbanks, Alaska 99775, U.S.A.*

Possible evidence for deep-seated magmatic activity beneath the northeastern Japan arc has been obtained from micro-earthquake observations. Tomographic inversions for *P*-wave velocity clearly delineate low-velocity zones, which are inclined to the west and continuously distributed from the uppermost mantle to the upper crust beneath active volcanoes. Exceptionally deep (22–40 km) micro-earthquakes are found at 10 locations around the low-velocity zones beneath active volcanoes. All these events have extremely low predominant frequencies, suggesting a close relation to the magmatic activity in this depth range. At shallower depths (8–15 km) distinct reflectors of *S*-waves from shallow events are found at five locations in or around the low-velocity zones. Their locations and reflection coefficients suggest that the reflectors are very thin magma bodies existing in the mid-crust. These observations shed some light on the state of magma at depths beneath volcanic arcs.

1. Introduction

Northeastern Japan is located at a typical subduction zone, where the oceanic Pacific plate subducts under the continental plate at an angle of about 30°. Many shallow earthquakes occur beneath the Pacific Ocean between the Japan trench and the Pacific coast, most of them originating at the boundary of the two converging plates. The shallow seismicity along the plate boundary is continuous with the deep seismic zone, which is within the subducted Pacific plate and attains a depth of about 650 km. The deep seismic zone, down to a depth of about 150 km, is composed of two thin planes which are parallel to each other and are 30–40 km apart. Very shallow earthquakes, confined to the upper crust, occur beneath the land area, although their activity is not very high. Active volcanoes are distributed in the land area, mainly along the volcanic front which runs through the middle of the land area almost parallel to the trench axis.

The seismic network of Tohoku University covers this tectonically active area. Seismic signals from high-gain short-period three-component seismic stations are centrally recorded on digital magnetic tape at the observation centre in Sendai by using telephone telemetry, which facilitates accurate hypocentre location of micro-earthquakes occurring in this area and enables us to accurately estimate the seismic velocity structure of the crust and upper mantle beneath this area. Detailed studies on seismic activity and on the crust and upper mantle structure, recently made by using data obtained from this network, show some evidence for magmatic activity in the crust and upper mantle. In the present paper we describe briefly the results of these studies on deep structure of active volcanoes beneath this area, the northeastern Japan arc, based on the seismic observations.

2. Low-velocity zones beneath active volcanoes

Many tomographic studies on local or regional scale have been carried out in various areas of the world since the pioneering works of seismic tomography by Aki & Lee (1976) and Aki *et al.* (1977). In subduction zones, the most extensive studies have been made in the Japan Islands. Hirahara *et al.* (1989) investigated the *P*-wave velocity structure beneath central Japan, and found low-velocity zones in the wedge mantle above the high-velocity Pacific plate subducted beneath this region. One of these low-velocity bodies coincides with an S-wave anisotropic body detected by Ando *et al.* (1983). Tomographic studies by Hasemi *et al.* (1984) and Obara *et al.* (1986) in northeastern Japan reveal *P*-wave low-velocity zones existing in the crust and in the wedge mantle just beneath active volcanoes.

The results of these studies show that tomographic studies play an important part for a deeper understanding of arc volcanism. In these studies, however, there still remain the following problems as in most of the conventional tomographic studies. It is known that there exist seismic velocity discontinuities, such as the Moho and Conrad discontinuities and the upper boundary of the subducted slab, in the crust and upper mantle beneath the northeastern Japan arc. However they do not take into consideration the effect of the complicated-shaped discontinuities or even their existence, which distorts the estimated tomographic images especially in the vicinity of the discontinuities. Another problem is that three-dimensional (3D) ray tracing is not used to calculate seismic ray paths and travel times. In very heterogeneous regions, such as subduction zones, calculated ray paths based on a simple 1D velocity model deviate considerably from the real paths in large hypocentral distances; this again seriously distorts the final tomographic images.

We developed a new tomographic method to solve the above-mentioned problems (Zhao *et al.* 1992). This method copes with a general velocity structure with complicated-shaped velocity discontinuities in the modelling space and with 3D variations in the seismic velocity in each layer bounded by the velocity discontinuities. An efficient 3D ray tracing algorithm is developed, which calculates ray paths and travel times very rapidly and accurately for *P* and *S* waves in a complicated velocity structure as described above. The present method can include in the inversion not only arrival time data of first *P* and *S* waves but also those of later phases such as reflected or converted waves at the velocity discontinuities. Details of the method are described in Zhao *et al.* (1992).

P-wave tomographic images of the crust and upper mantle beneath the northeastern Japan arc are obtained by applying the method to arrival time data of first *P* and *S* waves and *PS* and *SP* waves converted at the velocity discontinuities (Hasegawa *et al.* 1991; Zhao *et al.* 1992). The data used are those observed by the seismic networks of Tohoku University and several other national universities in Japan. This study has updated the tomographic images of the northeastern Japan arc by Hasemi *et al.* (1984) and Obara *et al.* (1986) by covering a wider area and by improving the resolution. The estimated 3D *P*-wave velocity structure is drawn with light and shade on vertical cross sections (figure 1), and on a horizontal plane (figure 2). Figure 1 shows three vertical cross sections of fractional *P*-wave velocity perturbations (in percent) along the lines AA', BB' and CC' in the inset map, which

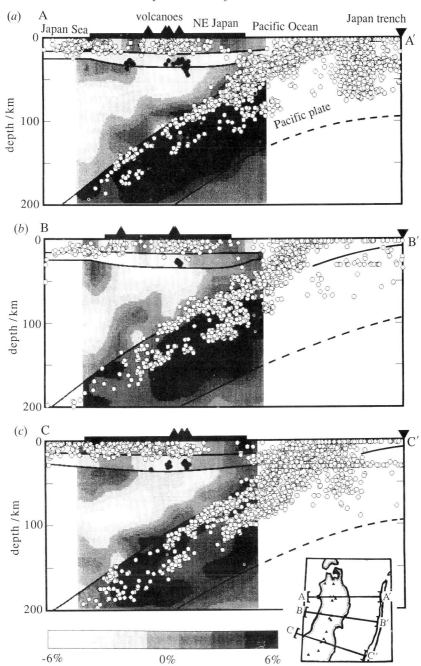

Figure 1. Vertical cross sections of fractional *P*-wave velocity perturbation (as %) along the lines (*a*) AA′, (*b*) BB′ and (*c*) CC′ in the insert map. Velocity perturbation is shown by the light and shade scale at the bottom. Open circles are micro-earthquakes within a 60 km width along each line located in 1987–90, and solid circles denote low-frequency micro-earthquakes located in 1975–90. The locations of the land area, trench axis and active volcanoes are shown on the top by a thick horizontal line, inverted solid triangles and solid triangles, respectively. The Conrad and Moho discontinuities and the top of the subducted slab, which are fixed in the inversion, are shown by thick lines. The estimated location of the bottom of the slab is also shown by a thick line.

are nearly perpendicular to the trench axis. Figure 2 shows the fractional *P*-wave velocity perturbations at a depth of 40 km. The velocity perturbation is from the mean value of estimated velocities at each depth, and the perturbation scale is from -6% to 6%. Light and shaded areas correspond to low and high velocities, respectively. In figure 1 *a–c*, shown by circles are micro-earthquakes within a 60 km width along each line located by the seismic network of Tohoku University. Also shown by thick curves are the locations of the Conrad and Moho discontinuities and the upper boundary of the subducted Pacific plate, which are fixed in the inversion.

As seen from figure 1, a high-velocity zone corresponding to the subducted Pacific plate is clearly delineated in all the three vertical sections. The lower boundary of this high-velocity zone can be recognized. The location of the lower boundary presently estimated is shown by a solid (and broken) curve in the figure. The thickness of the high-velocity Pacific plate is estimated as 80–90 km. This is in close agreement with an estimation by Umino *et al.* (1991), who detected a reflected and *S* to *P* converted phase at the bottom of the subducted Pacific plate in seismograms of intermediate-depth events occurring within the plate. They estimated the thickness of the plate to be *ca.* 85 km based on arrival time analyses of this phase. Figure 1 also shows that the double-planed deep seismic zone (Hasegawa *et al.* 1978) is located within the upper half of the high-velocity Pacific plate.

It is obvious from figure 2 that most of the active volcanoes (solid triangles) are located just above the low-velocity zones in the uppermost mantle. The low-velocity zones are also distributed in the crust just beneath active volcanoes; this is partly seen in figure 1. The low-velocity zones just beneath active volcanoes (solid triangles in figure 1) in the crust and in the uppermost mantle dip to the west in the mantle wedge and extend to a depth of about 150 km. They are approximately parallel to the dip of the subducted Pacific plate. This feature is common to all the low-velocity zones in the crust and in the mantle wedge shown in figure 1 *a–c*.

3. Cut-off depth for shallow inland seismicity

In the northeastern Japan arc very shallow earthquakes occur beneath the land area, although the activity of these crustal events is secondary to the major activity along the plate boundary beneath the Pacific Ocean. These shallow events are confined mainly to the so-called 'granitic layer' (Takagi *et al.* 1977). The accuracy of hypocentre locations of shallow events beneath the land area by the seismic network of Tohoku University has been much improved in the past three years. A detailed seismicity study made recently reveals that the cut-off depth for this shallow seismicity does not coincide with the Conrad depth (*ca.* 18 km beneath the northeastern Japan arc (Zhao *et al.* 1990)), but is slightly shallower than that. Figure 3 shows focal depth distribution of shallow micro-earthquakes accurately located in the land area of the northeastern Japan arc. Most of the shallow events (more than 98%) are found to be shallower than 15 km. The cut-off depth for this shallow seismicity, which is sharply delimited, can be interpreted as a brittle to ductile transition or a stick–slip to stable-sliding transition due to increasing temperature with depth (Brace & Byerlee 1970; Meissner & Strehlau 1982; Sibson 1982; Tse & Rice 1986). The upper 15 km of the crust forms a brittle seismogenic zone.

If examined in more detail, the cut-off depth for this shallow seismicity is found to vary with the location. Regional variations of the cut-off depth can be seen on figure 4. This figure shows a vertical cross section of shallow micro-earthquakes

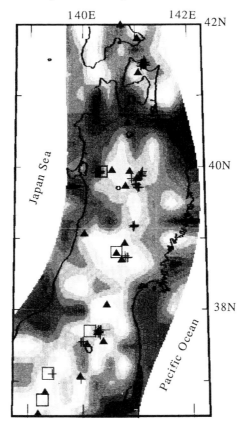

Figure 2. Fractional *P*-wave velocity perturbation (as %) at a depth of 40 km. Velocity perturbation is shown by the same light and shade scale as in figure 1. Solid triangles, crosses and open squares denote the locations of active volcanoes, low-frequency micro-earthquakes and *S*-wave reflectors, respectively.

within a 60 km width along the volcanic front accurately located in the central part of the northeastern Japan arc. Focal depths of all these events (except five anomalously deep events with low predominant frequencies) are shallower than about 15 km and their cut-off depth changes with location, becoming shallow beneath active volcanoes (solid triangles). The 3D *P*-wave velocity structure already described in §2 shows lower velocities in the crust beneath these active volcanoes than those beneath the other areas. The local elevation of the cut-off depth for the shallow seismicity near active volcanoes can be explained by the local elevation of the depth to the brittle to ductile (or stick–slip to stable-sliding) transition zone due to a higher temperature in the crust under those places.

4. Deep low-frequency micro-earthquakes beneath active volcanoes

Focal depth distributions of shallow events in the land area (figure 3) show that most events (more than 98%) occur in the upper 15 km of the crust, which forms a brittle seismogenic zone. However, exceptionally deep micro-earthquakes, well below the base of the brittle seismogenic zone, actually occur in the land area of the northeastern Japan arc, although the frequency of their occurrence is very low (0.97% of shallow events in the land area). These deep events can be seen in figure

frequency (%)

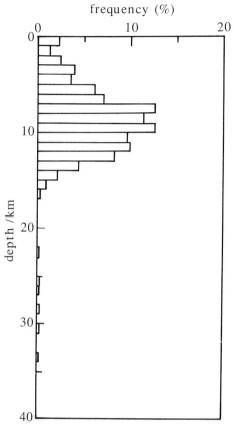

Figure 3. Depth-frequency distribution of shallow micro-earthquakes located in the land area of
the northeastern Japan arc in November 1988 through April 1990.

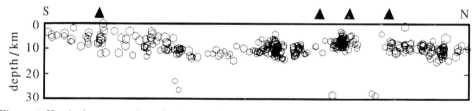

Figure 4. Vertical cross section of micro-earthquakes (open circles) within a 60 km width along the
volcanic front located in the central part of the northeastern Japan arc in November 1989 through
October 1991. Solid triangles on the top show the locations of active volcanoes.

3 as a distribution at depths greater than 22 km, which is distinctly isolated from the
main activity shallower than 17 km. Five of the deep events are also seen on the
vertical cross section in figure 4. We have so far found 151 deep events that occurred
at 10 locations in the northeastern Japan arc. A detailed investigation of these deep
events reveals that all of these events have the following anomalous features: (1)
their focal depths are anomalously deep (22–40 km); (2) they have extremely low
predominant frequencies (1.5–3.5 Hz) both for P and S waves; (3) their magnitudes
are at most 2.5; (4) they have very long duration times of oscillations compared with
events with normal focal depths; and (5) they occur around active volcanoes or
around the P-wave low-velocity zones.

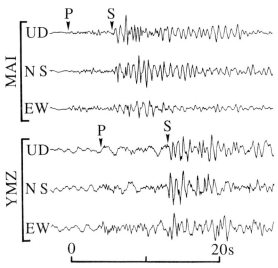

Figure 5. An example of three-component seismograms of a deep low-frequency micro-earthquake with magnitude of 1.8 that occurred beneath Hiuchidake volcano. Arrival times of direct *P* and *S* waves are denoted by *P* and *S*, respectively.

An example of three-component seismograms of a deep micro-earthquake recorded by short-period (1 s) seismometers is shown in figure 5. This shows seismograms of the deep event that occurred beneath Hiuchidake volcano in southwestern Fukushima Prefecture. The event has extremely low predominant frequencies both for *P* and *S* waves and very long duration times of oscillations.

All the 151 events detected at depths 22–40 km at 10 locations in the northeastern Japan arc are found to have extremely low predominant frequencies and long duration times. Earthquakes at depths 70–150 km, which occur in the deep seismic zone just under these low-frequency events, again have normal predominant frequencies both for *P* and *S* waves. Even if seismic waves from these earthquakes pass through the focal areas of the low-frequency events, their predominant frequencies are still normal. This means that the low predominant frequencies of these anomalously deep events are not caused by the path-effect.

Figure 6 shows the hypocentre distribution of the deep low-frequency events (open squares) that occurred at two locations near Kurikoma volcano (solid triangles) in southwestern Iwate Prefecture. Events with normal focal depths are also shown by open circles. All the events in this figure are those relocated by using the master-event method to improve the accuracy of hypocentre determinations. The low-frequency events shown by open squares are clearly isolated from the main activity of normal focal-depth events. This is common to all the low-frequency events that occurred at the other eight locations in the northeastern Japan arc, as can be estimated also from the focal depth distribution of shallow events in the land area (figure 3).

On the vertical cross sections in figure 1, the low-frequency events within a 60 km width along the lines AA′, BB′ and CC′ in the inset map are plotted by solid circles. Epicentres of all the low-frequency events detected at 10 locations are plotted by crosses on the map of *P*-wave velocity distribution in figure 2. The low-frequency micro-earthquakes are located approximately beneath active volcanoes (solid triangles) or around the *P*-wave low-velocity zones (light areas).

Phil. Trans. R. Soc. Lond. A (1993)

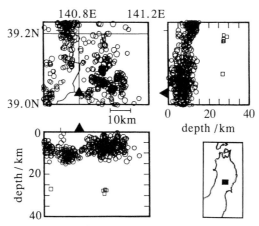

Figure 6. Hypocentre distribution of shallow micro-earthquakes (open circles) located near Kurikoma volcano in 1975–90. Low-frequency micro-earthquakes are shown by open squares. Solid triangles denote the location of Kurikoma volcano.

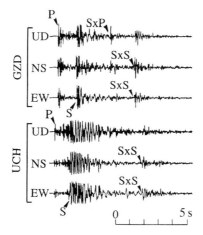

Figure 7. An example of three-component seismograms of a shallow event that occurred near Nikko-Shirane volcano. *P*- and *S*-wave first arrivals are denoted by *P* and *S*. Later arrivals are clearly seen and are indicated by *SxS* and *SxP*.

We have tried to estimate focal mechanisms of the low-frequency events by using initial motions of *P* waves, although the number of polarity data and their coverage on the focal sphere are not sufficient because of their small magnitudes. A relatively large event that occurred beneath Osoreyama volcano in northern Aomori Prefecture has a distribution of *P*-wave initial motions on the focal sphere inconsistent with a double-couple mechanism. A preliminary estimation by a moment-tensor inversion using first *P*- and *S*-wave amplitudes for this event also suggests a non-double-couple rather than a normal double-couple mechanism (Kosuga *et al.* 1991). The close locations of these anomalous events to active volcanoes and other features described above suggest that they are generated by magmatic activity, such as the rapid movement of magma, in the depth range from 22 to 40 km.

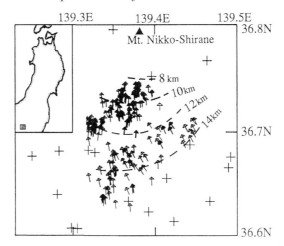

Figure 8. Geometry of the *S*-wave reflector beneath Nikko-Shirane volcano. The estimated depth to the reflector is shown by broken contours. Arrows denote the reflection points of observed reflected waves and the local upgrade directions of the reflector. Locations of observation stations temporarily deployed are shown by crosses.

5. Mid-crustal magma bodies estimated from *S*-wave reflections

Distinct *S*-wave reflections from shallow earthquakes have been found at five locations of the northeastern Japan arc. Figure 7 shows an example of three-component seismograms of a shallow micro-earthquake that occurred near Nikko-Shirane volcano in western Tochigi Prefecture. A sharp impulsive phase (denoted by '*SxS*' in the figure) following the direct *S* wave is recognized at two stations, which are located just above the hypocentre of this event. This unusual phase is most clearly defined on horizontal component seismograms and its amplitude is very large, in some cases being even larger than that of the direct *S* wave. Arrival time analysis indicates that this phase is an *S* wave reflected from a velocity discontinuity in the mid-crust (*SxS* phase). A reflected and *S* to *P* converted phase (*SxP* phase) at the same discontinuity is also detected as can be seen on the vertical component seismogram at one of the two stations shown in figure 7 (denoted by '*SxP*').

Figure 8 shows the estimated location of the *S*-wave reflector by analysing arrival times of the *SxS* phase observed by a dense seismic network temporarily deployed in this region (Matsumoto & Hasegawa 1991). The reflector has a shape similar to a section of a cone and is distributed over an area of 15×10 km^2 at depths ranging from 8 to 15 km. It becomes shallow at an angle of about $30°$ toward the north, in which direction Nikko-Shirane volcano (solid triangle) is located.

Distinct *SxS* and *SxP* phases from a reflector in the mid-crust, similar to the present case, were first detected and identified beneath the Rio Grande Rift near Socorro, New Mexico (Sanford *et al.* 1973). The large amplitude of the *SxS* phase and *SxP* to *SxS* amplitude ratios are explained by a large velocity contrast across a discontinuity underlain by low-rigidity material, such as a magma body (Sanford *et al.* 1973). Observed spectral ratios of *SxS* to direct *S* phases have three peaks in the frequency range from 3 to 20 Hz, which can be explained by a very thin (thickness of *ca.* 100 m) magma body containing low-rigidity (*P*- and *S*-wave velocities of *ca.* 3 and 1 km s^{-1}, respectively) material (Matsumoto & Hasegawa 1991). A very thin magma body is partly supported by the following observation. Direct *S* waves

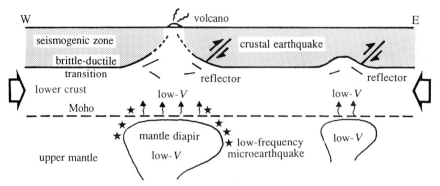

Figure 9. Schematic east–west cross section of the crust and upper mantle structure beneath the northeastern Japan arc.

observed at stations just above the magma body from earthquakes in the deep seismic zone right under it, which pass through the estimated magma body in the mid-crust, do not have spectra different from those at other stations. This indicates that S waves passing through the magma body are not attenuate significantly, suggesting that the thickness of the magma body is very thin, similar to that detected beneath the Rio Grande Rift, New Mexico (Ake & Sanford 1988).

Distinct S-wave reflectors in the mid-crust have been detected so far at five locations in the northeastern Japan arc (Mizoue et al. 1982; Horiuchi et al. 1988; Iwase et al. 1989; Hasegawa et al. 1991). Locations of the reflectors are plotted by large squares on the map of P-wave velocity distribution in figure 2. As is obvious from this figure, all the reflectors are located near active volcanoes (solid triangles) and/or in or around the low-velocity zones of P wave (light areas). Deep low-frequency micro-earthquakes (crosses) are distributed at greater depths (22–40 km) beneath the reflectors. These observations and the features of SxS phase described above support the evidence that the SxS and SxP phases observed are reflected phases from thin magma bodies in the mid-crust.

6. Discussion and conclusions

We infer the structure of the crust and upper mantle beneath the northeastern Japan arc schematically as shown in figure 9, by combining the observations described in the previous sections (Hasegawa et al. 1991). The low-velocity zones inclined to the west in the upper mantle perhaps correspond to ascending flow of subduction-induced convection in the mantle wedge (Toksöz & Bird 1977). The migration of mass in the ascending flow causes upwelling of hot mantle material, and thus produces low-velocity zones which form roots of arc volcanoes at depth. Low-velocity zones just under the Moho discontinuity are the manifestation of mantle diapirs, and their magmatic activity (such as the rapid movement of magma) may generate low-frequency micro-earthquakes around them.

Molten material rising from the uppermost mantle makes its appearance in the mid-crust as distinct S-wave reflectors which are supposed to be very thin magma bodies. They also raise the temperature of crustal materials around them, causing the local elevation of the base of the brittle seismogenic zone. The lower portion of the crust and the mantle wedge, below the base of the seismogenic zone, are governed by creep and flow, because of a large geothermal gradient beneath this volcanic arc, and they are incapable of supporting much stress (Shimamoto 1991). Most of the horizontal compressional stress as caused by subduction of the Pacific plate is supported by the upper 15 km of the crust which forms the brittle seismogenic zone. This situation causes stress concentration around places where the base of the seismogenic zone is locally elevated, and shallow crustal earthquakes occur in those places. Our interpretation seems to be supported in part by the fact that many large crustal earthquakes occur in or around the *P*-wave low-velocity zones (Hasegawa *et al.* 1991).

Spatial resolution of our *P*-wave tomographic inversion is 25–30 km (Zhao *et al.* 1992). The low-velocity zones in the uppermost mantle, which are supposed to be mantle diapirs, are just the images obtained with this resolution. Further investigation with much higher resolution would provide clearer images of them. Making use of reflected seismic waves is a powerful approach to the improvement of spatial resolution. Recently, in seismograms of shallow micro-earthquakes, we have detected a clear *S* wave reflected from a sharp velocity discontinuity in the uppermost mantle just beneath Osoreyama volcano in northern Aomori Prefecture (Hori & Hasegawa 1991). This distinct *S*-wave reflector is located over an area of 5×5 km^2 at depths of 44–48 km within a *P*-wave low-velocity body of the uppermost mantle, and becomes shallow toward the north-northeast at an angle of about 40°. On the extension of it, low-frequency micro-earthquakes are distributed at depths of 25–35 km. Seismic observations for obtaining more detailed structure of mantle diapirs will make an important contribution to a better understanding of arc volcanism.

Finally, the deep structure of active volcanoes beneath the northeastern Japan arc obtained to date from seismic observations is summarized as follows. Clear *P*-wave low-velocity zones are found in the crust and in the mantle wedge. The low-velocity zones in the crust and the uppermost mantle just beneath active volcanoes are inclined to the west in the mantle wedge and attain a depth of about 150 km. Around the low-velocity zones there occur anomalously low-frequency micro-earthquakes at depths of 22–45 km, perhaps caused by magmatic activity in this depth range. Very thin magma bodies are found in or around the low-velocity zones at depths of 8–15 km in the mid-crust. A brittle seismogenic zone, corresponding to the upper 15 km of the crust, changes its thickness with the location, and its base is locally elevated just beneath active volcanoes.

References

Ake, J. P. & Sanford, A. R. 1988 New evidence for the existence and internal structure of a thin layer of magma at mid-crustal depths near Socorro, New Mexico. *Bull. Seism. Soc. Am.* **78**, 1335–1359.

Aki, K. & Lee, W. H. K. 1976 Determination of three-dimensional velocity anomalies under a seismic array using first *P* arrival times from local earthquakes. 1. A homogeneous initial model. *J. geophys. Res.* **81**, 4381–4399.

Aki, K., Christoffersson, A. & Husebye, E. S. 1977 Determination of the three-dimensional seismic structure of the lithosphere. *J. geophys. Res.* **82**, 277–296.

Ando, M., Ishikawa, Y. & Yamazaki, F. 1983 Shear wave polarization anisotropy in the upper mantle beneath Honshu, Japan. *J. geophys. Res.* **88**, 5850–5864.

Brace, W. F. & Byerlee, J. D. 1970 California earthquakes: why only shallow focus? *Science, Wash.* **168**, 1573–1575.

Hasegawa, A., Umino, N. & Takagi, A. 1978 Double-planed deep seismic zone and upper mantle structure in the northeastern Japan arc. *Geophys. Jl R. astr. Soc.* **54**, 281–296.

Hasegawa, A., Zhao, D., Hori, S., Yamamoto, A. & Horiuchi, S. 1991 Deep structure of the
    northeastern Japan arc and its relationship to seismic and volcanic activity. *Nature, Lond.* **352**,
    683–689.

Hasemi, A. H., Ishii, H. & Takagi, A. 1984 Fine structure beneath the Tohoku District,
    northeastern Japan arc, as derived by an inversion of *P*-wave arrival times from local
    earthquakes. *Tectonophys.* **101**, 245–265.

Hori, S. & Hasegawa, A. 1991 Location of a mid-crustal magma body beneath Mt. Moriyoshi,
    northern Akita Prefecture, as estimated from reflected *SxS* phases. *J. Seism. Soc. Japan* **44**,
    39–48.

Hori, S. & Hasegawa, A. 1991 Anomalous *S*-wave reflector in the upper mantle beneath
    Osoreyama volcano. *Programme and Abstracts of Seism. Soc. Japan* no. 2, 208.

Horiuchi, S., Hasegawa, A., Takagi, A., Ito, A., Suzuki, M. & Kameyama, H. 1988 Mapping of
    a melting zone near Mt. Nikko-Shirane in northern Kanto, Japan, as inferred from *SxP* and
    SxS reflections. *Tohoku Geophys. J.* **31**, 43–55.

Hirahara, K., Ikami, A., Ishida, M. & Mikumo, T. 1989 Three-dimensional *P*-wave velocity
    structure beneath central Japan: low-velocity bodies in the wedge portion of the upper mantle
    above high-velocity subducting plates. *Tectonophys.* **163**, 63–73.

Iwase, R., Urabe, S., Katsumata, K., Moriya, M., Nakamura, I. & Mizoue, M. 1989 Mid-crustal
    magma body in southwestern Fukushima Prefecture detected by reflected waves from
    microearthquakes. *Programme and Abstracts of Seism. Soc. Japan* no. 1, 185.

Kosuga, M., Michinaka, M. & Hasegawa, A. 1991 Focal mechanism of low frequency earthquakes
    in the northeastern Japan arc. 1 Case study of the earthquakes in Shimokita region. *Abstracts for
    the 1991 Japan Earth and Planetary Science Joint Meeting*, 105.

Matsumoto, S. & Hasegawa, A. 1991 Characteristics of mid-crustal *S*-wave reflector in Nikko-
    Ashio region. *Programme and Abstracts of Seism. Soc. Japan*, 204.

Meissner, R. & Strehlau, J. 1982 Limits of stresses in continental crusts and their relation to the
    depth-frequency distribution of shallow earthquakes. *Tectonics* **1**, 73–89.

Mizoue, M., Nakamura, I. & Yokota, T. 1982 Mapping of an unusual crustal discontinuity by
    microearthquake reflections in the earthquake swarm area near Ashio, northwestern part of
    Tochigi Prefecture, central Japan. *Bull. Earthquake Res. Inst.* **57**, 653–686.

Obara, K., Hasegawa, A. & Takagi, A. 1986 Three-dimensional *P* and *S* wave velocity structure
    beneath the northeastern Japan arc. *J. Seism. Soc. Japan* **39**, 201–215.

Sanford, A. R., Alptekin, O. & Toppozada, T. R. 1973 Use of reflection phases on microearthquake
    seismograms to map an unusual discontinuity beneath the Rio Grande Rift. *Bull. Seism. Soc.
    Am.* **63**, 2021–2034.

Sibson, R. H. 1982 Fault zone models, heat flow, and the depth distribution of earthquakes in the
    continental crust of the United States. *Bull. Seism. Soc. Am.* **72**, 151–163.

Shimamoto, T. 1991 Rheology of rocks and plate tectonics – from rigid plates to deformable
    plates. *Comprehensive Rock Engng.* (Submitted.)

Takagi, A., Hasegawa, A. & Umino, N. 1977 Seismic activity in the northeastern Japan arc.
    J. Phys. Earth **25**, 95–104.

Toksöz, M. N. & Bird, P. 1977 Formation and evolution of marginal basins and continental
    plateaus. In *Island arcs, deep sea trenches and back-arc basins* (ed. M. Talwani & W. C. Pitman).
    AGU.

Tse, S. T. & Rice, J. R. 1986 Crustal earthquake instability in relation to the depth variation of
    frictional slip properties. *J. geophys. Res.* **91**, 9452–9472.

Zhao, D. & Hasegawa, A. 1992 *P*-wave tomographic imaging of the crust and upper mantle
    beneath the Japan Islands. *J. geophys. Res.* (In the press.)

Zhao, D. & Hasegawa, A. & Horiuchi, S. 1992 Tomographic imaging of *P* and *S* wave velocity
    structure beneath northeastern Japan. *J. geophys. Res.* (In the press.)

Zhao, D., Horiuchi, S. & Hasegawa, A. 1990 3-D seismic velocity structure of the crust and the
    uppermost mantle in the northeastern Japan arc. *Tectonophys.* **181**, 135–149.

Trace element fractionation processes in the generation of island arc basalts

By C. J. Hawkesworth, K. Gallagher, J. M. Hergt†
and F. McDermott

Department of Earth Sciences, The Open University, Milton Keynes MK7 6AA, U.K.

Subduction-related magmas are characterized by distinctive minor and trace element ratios which are widely attributed to the introduction of a hydrous component from the subducted crust. Island arc rocks may usefully be subdivided into high and low Ce/Yb groups, and the latter are characterized by relatively restricted radiogenic isotope ratios. In general, high LIL/HFSE ratios are best developed in low HFSE rocks, and the variation in LILE is less than that in HFSE. A local equilibrium model is developed in which the distinctive minor and trace element feature of arc rocks are the result of fluid percolation in the mantle wedge. Peridotite/fluid distribution coefficients are inferred to vary systematically with ionic radius in the range 69–167×10^{-12} m. However, in practice the calculated olivine/fluid partition coefficients are too high to develop an arc signature in the wedge peridotite in reasonable timescales, and for acceptable fluxes from the slab. The available geochemical data would suggest that realistic distribution coefficients are 2–3 orders of magnitude less than those presently available from experimental data, presumably because the fluid compositions are different, or that local equilibrium is not appropriate. Average compositions from the low Ce/Yb arc suites exhibit a positive correlation between Ce/Sm, but not K/Sm, and crustal thickness. It is argued that the degree of melting varies with crustal thickness, but not in any simple way with the magnitude of the fluid contribution. The observed range in Ce/Sm in the low Ce/Yb rocks is consistent with 3–18% melting of slightly LREE depleted source rocks.

1. Introduction

Magma generation along destructive plate margins differs significantly from that in other tectonic settings. First, subduction-related magmas have distinctive minor and trace element compositions; second, considerable volumes of magma are generated despite the depression of mantle isotherms caused by the subduction of cold oceanic crust; and third, partial melting appears to be initiated by the introduction of water and depression of the peridotite solidus, rather than by diapiric upwelling and decompression. Recent studies have reviewed the major element variations in arc magmas, and sought to model them in the light of the limited experimental data available (Plank & Langmuir 1988; Davies & Bickle 1991). This contribution therefore discusses some of the minor and trace element features of magmas generated above subduction zones, and evaluates the extent to which they are consistent with the major element models.

† Present address: Research School of Earth Sciences, The Australian National University, GPO Box 4, Canberra, ACT 2601, Australia.

Plank & Langmuir (1988) concluded that both the major element abundances, and the inferred degrees of melting, for primitive mid-ocean ridge (MOR) and arc basalts were surprisingly similar. Whatever the causes of melting in the two environments, this conclusion was taken as further evidence that in most cases it is the mantle wedge which melts beneath arcs, rather than the subducted slab. In detail, Plank & Langmuir (1988) re-emphasized the near constancy of the depth to the Benioff Zone beneath volcanic arcs, at *ca.* 110 km, which suggests that arc magmatism is linked to a pressure sensitive reaction, such as the breakdown of amphibole. They also demonstrated that Na_2O at 6% MgO increased, and CaO at 6% MgO decreased significantly with increasing thickness of the arc crust. Variations of other chemical features, such as average SiO_2, with crustal thickness have been described previously (Coulon & Thorpe 1981; Leeman 1983), but for the most part attributed to increased fractional crystallization and crustal contamination in areas of thicker crust. However, the variations in Na_2O and CaO at 6% MgO are difficult to explain by any such shallow level processes, rather they are consistent with a model in which the average degree of melting depends on the thickness of the melting column (Plank & Langmuir 1988).

2. Minor and trace elements, and radiogenic isotopes

Several minor and trace element features of subduction-related rocks are consistently different from those of MORB and OIB, and in particular they are characterized by high LIL/HFS element ratios, and relatively low Nb, Ta, and perhaps Ti abundances (Pearce 1982). In contrast, the majority of destructive plate margin rocks have Nd, Sr and Pb isotope ratios which are broadly similar to those of OIB (Morris & Hart 1983; Hawkesworth *et al.* 1991*a*), and to those of MORB which have more enriched isotope signatures (higher $^{87}Sr/^{86}Sr$ and lower $^{143}Nd/^{144}Nd$). Some subduction related rocks exhibit steep arrays on Pb isotope diagrams (Kay *et al.* 1978; Woodhead & Fraser 1985; White & Dupré 1986), consistent with the introduction of radiogenic Pb from subducted sediments, and more recently it has been demonstrated that young arc rocks often exhibit high ^{10}Be which requires a contribution from young (less than 5 Ma) sedimentary material, presumably in the subducted slab (Tera *et al.* 1986; Morris *et al* 1990). In general, however, the isotope data on subduction-related rocks suggest that the contributions from sediments and altered oceanic crust in the subducted slab are typically much less than estimates of the 'subduction component' calculated on the basis of minor and trace element variations (Pearce 1983; Hawkesworth *et al.* 1991*a*).

The rare earth elements (REE) are widely used in models of petrogenesis, both because they are a geochemically coherent group and because they include the radioactive decay scheme of ^{147}Sm to ^{143}Nd. A key step is to identify rock suites which may have been derived from similar source regions, and Hawkesworth *et al.* (1991*a*) pointed out that the REE results from destructive plate margin rocks broadly fall into two groups. In one, Ce and Yb vary together, and in the other the Ce contents are much higher and the Yb contents are similar. Such differences require major differences in either the REE profiles of the source rocks for the two groups, and/or in the bulk distribution coefficients and the degree of melting. Rocks in the high Ce/Yb group have a much greater range in Nd, Sr and Pb isotope ratios, and so the simplest interpretation is that they contain at least a contribution from material which in many cases was both old (i.e. old enough to have developed

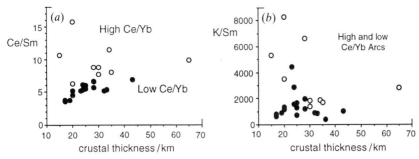

Figure 1. Variations in the average Ce/Sm and K/Sm ratios of arc rocks with crustal thickness. The low Ce/Yb arcs (●) include data from the Aleutians, Andes (SVZ), Central Americas, Fiji, Halmahera, Honshu, Kermadecs, Kuriles, L. Antilles, Papua, Marianas, NE Sulawesi, New Britain, Tonga and Vanuatu: the high Ce/Yb arcs (○) include data from the Aeolian Islands, Andes (CVZ), Costa Rica, Grenada, Java, Papua, Muriah, the Philippines and Sunda (references in Hawkesworth *et al.* 1991*b*). Average MgO contents = 4.9–7.8%, and there is no systematic variation between Ce/Sm and MgO.

different isotope ratios), and LREE enriched. Such features have been attributed both to subducted sediments and to trace element enriched source regions in the mantle wedge.

The dilemma facing any analysis of partial melting processes from minor and trace element data is the minor and trace element composition of the source, and whether that can be inferred from chemical trends in the more primitive rock samples, or it must be assumed. McKenzie & O'Nions (1991) used inverse theory to calculate the melt fractions with depth required to produce the REE concentrations in magmatic rocks from different tectonic settings. They chose to average selected data-sets, and to assume the REE contents of the magma source regions. Such calculations reaffirm the sensitivity of REE profiles to the presence or absence of residual garnet, and they encouraged McKenzie & O'Nions to invoke residual amphibole in the source of island arc magmas. However, they also indicate that for ratios of LREE, such as Ce/Sm, the rate of change with degree of melting is similar for garnet-free and garnet-bearing assemblages. Consequently Ce/Sm ratios may be used crudely to evaluate differences in the average degree of melting irrespective of the proportion of melting that took place in the garnet and spinel stability fields.

To evaluate the extent to which REE and other trace element ratios vary with crustal thickness, as demonstrated for Na_2O and CaO at 6% MgO (Plank & Langmuir 1988), average values have been compiled for arc rocks with less than 53% SiO_2. Within the low Ce/Yb group of arcs, average Ce/Sm increases from 3.6 to 6.9 as the crustal thickness increases from *ca.* 18 km at Tonga and the Northern Lesser Antilles to *ca.* 43 km in the Southern Andes. Similarly there is a broad positive correlation between average Ce/Sm and Na_2O/CaO. However, the high Ce/Yb arcs show much more scatter, and many of them are predictably displaced to higher Ce/Sm for any particular crustal thickness (figure 1*a*). Since the high Ce/Yb suites tend to have lower $^{143}Nd/^{144}Nd$, their higher Ce/Sm ratios are inferred, at least in part, to reflect their different source compositions. However, the variation in Ce/Sm within the low Ce/Yb rocks is consistent with different degrees of partial melting, with the smaller degrees of partial melting occurring in areas of thicker crust.

Fractional crystallization can affect REE ratios even in rocks with less than 53% SiO_2, and so it might be argued that the increase in Ce/Sm with crustal thickness in figure 1 is a function of more fractional crystallization in areas of thicker crust. However, using the D values in table 1, 50% clinopyroxene fractionation only changes Ce/Sm in the residual magma by 10%. Amphibole also fractionates REE ratios, but in this case Sm/Yb decreases as Ce/Sm increases, whereas in the rocks studied here there is a positive correlation between Ce/Sm and Sm/Yb. Thus it would appear that the differences between the average Ce/Sm ratios have not been changed significantly by fractional crystallization processes. Rather, if it is simply assumed that the low Ce/Yb rocks were derived from source regions with similar REE ratios, the observed range in Ce/Sm can be generated by *ca.* 3–18% fractional melting of a slightly LREE depleted source, using the distribution coefficients in table 1.

The distinctive relatively high LILE contents of arc rocks can be illustrated by the variations in K/Sm and, in contrast to Ce/Sm, there is no systematic variation of K/Sm with crustal thickness (figure 1*b*). The simplest interpretation is that the variation in K/Sm does not primarily reflect partial melting processes. Minor and trace element variations in arc basalts indicate that if the high K/Sm and high LIL/HFSE ratios were due to partial melting processes, for example, then $D_{Sm} \gg D_K$ and $D_{Ce} > D_{Nb}$. Many authors have concluded that this is unlikely, and that the relatively high LILE contents of subduction-related rocks are instead due to scavenging by hydrous fluids, from the slab and/or the mantle wedge (Pearce 1982; Arculus & Powell 1986; Tatsumi *et al.* 1986).

3. Melt and fluid percolation

Beneath volcanic arcs, hydrous fluids from the subducted slab, and subsequently partial melts, must migrate through mantle which is being dragged down by convection. During this fluid movement, trace and minor elements may be fractionated because of their differing effective transport velocities relative to the solid matrix (McKenzie 1984; Navon & Stolper 1987; McKenzie & O'Nions 1991). In the one-dimensional problem, assuming instantaneous local equilibrium between the fluid and the matrix during percolation, the effective velocity for an element *i* is

$$U_e^i = FU_f \qquad (1)$$

with $F = 1/(1+D^i/N)$ and U_f is the fluid velocity relative to a static matrix, N is the fluid/solid mass ratio and D^i is the solid/fluid partition coefficient. Thus, provided the initial fluid composition is different from that in equilibrium with the matrix, all minor and trace elements will move more slowly than the bulk fluid. In the absence of diffusion and dispersion, a constant flux of fluid of a given composition results in concentration fronts for individual elements. For the *i*th element, moving with velocity U_e^i, the fluid ahead of the front is in equilibrium with the pre-fluid element concentration in the matrix, while the fluid behind the front maintains its original composition, with the relative concentrations determined by the appropriate partition coefficient. In a subduction zone, where solid material in the wedge overlying the slab is moving down with velocity U_s, then, in steady state, a concentration front, C_f, lies along a direction given by (Navon & Stolper 1987)

$$(\partial z/\partial x)_{C_f} = (FU_f - U_s \sin \alpha)/U_s \cos \alpha, \qquad (2)$$

where α is the dip of the subduction zone from the horizontal. Clearly, if $F < (U_s/U_f)$ $\sin \alpha$ then $(\partial z/\partial x)_{C_f} < 0$, and the concentration front for an element with such a value

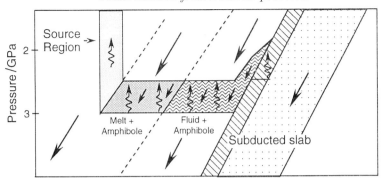

Figure 2. Schematic model for the release of hydrous fluids from the subducted slab and lateral migration to the zone of melt generation (after Davies & Bickle 1991). The fluids traverse laterally by a combination of vertical movement as a fluid, and downward movement fixed in amphibole carried down by induced mantle flow (straight arrows). In the preferred model, element fractionation may result from the percolation of first a hydrous fluid, and subsequently a partial melt through a peridotite matrix, and then during partial melting in the magma source regions.

Table 1. *Selected mineral/melt distribution coefficients used in this study*

	olivine	opx	cpx	spinel	garnet	amphibole
Pb	10^{-4}	1.3×10^{-3}	10^{-2}	5×10^{-4}	5×10^{-4}	10^{-1}
Rb	1.8×10^{-4}	6×10^{-4}	1.1×10^{-2}	5×10^{-4}	7×10^{-4}	2×10^{-1}
Ba	10^{-4}	6×10^{-4}	10^{-2}	5×10^{-4}	7×10^{-4}	2.5×10^{-1}
K	10^{-1}	6×10^{-4}	10^{-2}	5×10^{-4}	7×10^{-4}	8×10^{-1}
Th	1.1×10^{-4}	10^{-4}	2.6×10^{-4}	10^{-3}	10^{-4}	7×10^{-2}
U	10^{-4}	10^{-4}	3.6×10^{-4}	10^{-3}	10^{-4}	9×10^{-2}
Nb	10^{-2}	10^{-2}	5×10^{-2}	10^{-2}	5×10^{-3}	4×10^{-1}
La	4×10^{-4}	2×10^{-3}	5.4×10^{-2}	10^{-2}	10^{-2}	1.7×10^{-1}
Ce	5×10^{-4}	3×10^{-3}	9.8×10^{-2}	10^{-2}	2.1×10^{-2}	2.6×10^{-1}
Sr	1.9×10^{-4}	7×10^{-3}	6.7×10^{-2}	5×10^{-4}	1.1×10^{-3}	1.2×10^{-1}
Nd	10^{-3}	6.8×10^{-3}	2.1×10^{-1}	10^{-2}	8.7×10^{-2}	4.4×10^{-1}
Sm	1.3×10^{-3}	10^{-2}	2.6×10^{-1}	10^{-2}	2.17×10^{-1}	7.6×10^{-1}
Zr	4×10^{-3}	3×10^{-2}	2×10^{-2}	5×10^{-2}	2.5×10^{-1}	5×10^{-1}
Ti	6×10^{-3}	2.4×10^{-2}	10^{-1}	4.8×10^{-2}	10^{-1}	6.9×10^{-1}
Y	4×10^{-4}	9×10^{-2}	7×10^{-1}	10^{-2}	3.8	6×10^{-1}
Yb	1.5×10^{-3}	4.9×10^{-2}	2.8×10^{-1}	10^{-2}	4.03	5.9×10^{-1}

Peridotite modal proportions
(From McKenzie & O'Nions 1992.)

bulk rock	olivine	opx	cpx	gnt/sp/amp
spinel peridotite	0.578	0.270	0.119	0.033
garnet peridotite	0.598	0.211	0.076	0.115
amphibole peridotite	0.599	0.247	0.038	0.116

of F (and hence D) will move downwards with the wedge (figure 2). This analysis neglects the effects of disequilibrium, dispersion and magma mixing before eruption, all of which will tend to reduce the magnitude of the element fractionation and introduce smoother variations in element concentrations with time. However, the simple approach is useful for understanding the relative fractionation of elements by percolation for different fluids and partial melts.

Figure 2 is modified after Davies & Bickle (1991), who developed a physical model

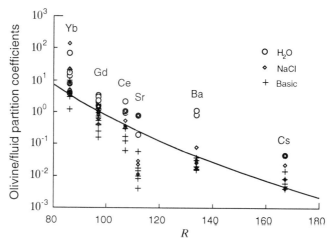

Figure 3. The olivine/fluid partition coefficients for Yb, Gd, Ce, Sr, Ba and Cs measured by Brenan & Watson (1991) plotted against ionic radius (R). The best fit curve is used to estimate partition coefficients for other minor and trace elements with R in the range $69–167 \times 10^{-12}$ m, using values of R from Henderson (1981).

for the volume and composition of melt produced by hydrous fluxing above subduction zones. Hydrous fluids, released into the mantle by dehydration of the subducted slab, traverse the wedge horizontally by a combination of vertical movement as a fluid, and downward movement fixed in amphibole carried down by the induced mantle flow. In such a model, minor and trace elements may be fractionated first by the movement of hydrous fluids, and second of partial melts through the mantle matrix.

Here we wish to evaluate the element fractionation effects of fluid percolation and to do so it is necessary to estimate appropriate partition coefficients. McKenzie & O'Nions (1991) recently compiled mineral/melt partition coefficients for the REEs and some minor elements. These data have been supplemented for other elements and the values used in this work are summarized in table 1. Solid/aqueous fluid partition coefficients are less well known (Eggler 1987), although Brenan & Watson (1991) have recently reported olivine/fluid partition coefficients for several elements and various fluid compositions. Their results indicate that the presence of dissolved salts, in particular carbonates and hydroxides, in the fluid can reduce the partition coefficient relative to pure water by up to two orders of magnitude. Additionally, Tatsumi *et al.* (1986) examined the relative mobility of incompatible elements in hydrous fluids and showed that mobility increases with increasing ionic radius (R), at least in the range for Nb to Cs ($R = 69$ to 167×10^{-12} m). This suggests that there is a relation between R and the solid/fluid partition coefficient, and figure 3 illustrates the olivine/fluid partition coefficients of Brenan & Watson (1991) as a function of ionic radius. As would be expected from the mobility arguments, the partition coefficients decrease with increasing radius, and this relationship has been approximated by a function of the form:

$$Ln(D_{o/f}) = a + b * Ln(R) \tag{3}$$

with $a = 45.63$, $b = -9.97$, and R in $m \times 10^{-12}$. This relationship may be used to predict olivine/fluid partition coefficients for other incompatible elements and, from these and the mineral/melt values mentioned above, and assuming equilibrium

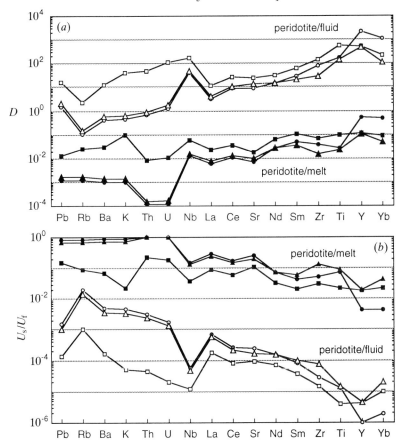

Figure 4. (*a*) Peridotite/fluid partition coefficients calculated from the olivine/fluid partition coefficients of Brenan & Watson (1991), as illustrated in figure 3, compared with the peridotite/melt partition coefficients based on the mineral/melt coefficients in table 1. (*b*) The solid/fluid velocity ratio U_s/U_f, for both melts and hydrous fluids, for which equation (2) with $\alpha = 45°$ is equal to zero. For velocity ratios greater than this critical value elements are expected to be carried down by convection in the wedge, and for lower velocity ratios elements will migrate at some angle above the horizontal (see also figure 2). □, ■, Amphibole peridotite; △, ▲, spinel peridotite; ○, ●, garnet peridotite.

between bulk rock melt and fluid we have estimated bulk rock/fluid partition coefficients according to

$$D_{\mathrm{br/f}} = \frac{D_{\mathrm{o/f}}}{D_{\mathrm{o/m}}} \sum_{j=1}^{N} D_{j\,\mathrm{m}} F_{\mathrm{o}}^{j} = \frac{D_{\mathrm{o/f}}}{D_{\mathrm{o/m}}} D_{\mathrm{br/m}}, \tag{4}$$

where N is the number of mineral phases, $D_{j\,\mathrm{m}}$ and F_{o}^{j} are the mineral/melt partition coefficient and fraction of the jth phase respectively. Figure 4*a* summarizes the calculated bulk rock/melt and bulk rock/fluid partition coefficients for spinel periodotite and amphibole peridotite using the modal proportions from McKenzie & O'Nions (1991). The bulk rock/fluid values are high and this may in part be because all of Brenan & Watson's experimental data were used to constrain $D_{\mathrm{o/f}}$ as a function of R, even though the results for fluids containing dissolved salts alone might be more representative of the real situation. However, the *relative* fractionation between elements should be unaffected by the choice of absolute values for the partition coefficients. We also investigated the correlation between the experimental values of Brenan & Watson (1991) and the ionic charge/radius ratio (Z/R) but could not obtain a significantly better fit to the data. The trends in the calculated partition coefficients using Z/R are broadly similar to those in figure 4*b*, although the range of values is greater, with D_{HFSE} being 2–3 orders of magnitude greater than shown in figure 4*b*, while D_{LILE} being up to a factor of five lower.

Figure 4*b* shows the solid/fluid velocity ratio, U_s/U_f, for both melt and hydrous fluid, for which equation (2), with $\alpha = 45°$, is equal to zero. For velocity ratios greater than this critical value, a given element, travelling with a lower effective velocity than the fluid, will have a sub-horizontal velocity vector, and hence is expected to be carried down by convection in the mantle wedge. At smaller values of U_s/U_f elements will migrate at some angle above the horizontal, and are more likely to arrive at the zone of melt generation. As the dip of the subduction zone decreases, the critical velocity ratio increases, so that more compatible elements may reach the melt generation zone for shallow subduction angles. The velocity ratios for the aqueous fluid are considerably lower than those for melt, because the inferred solid/fluid partition coefficients are much higher. However, as shown by the relevant form of Darcy's law at constant porosity, the velocity of aqueous fluid relative to the matrix is expected to be two or three orders of magnitude higher than that for melt because of the aqueous fluid's lower density and viscosity.

The critical U_s/U_f ratio for different elements largely depends on the differences in their estimated distribution coefficients. Thus for most elements the fluid velocities required for an element to migrate horizontally are significantly greater for amphibole than for spinel peridotite. In detail, it is noticeable that the LILE, and especially U, Th and K, are less enriched relative to the LREEs in the presence of amphibole, and for both peridotites Nb is predicted to be relatively more depleted in the aqueous fluid than in a partial melt.

The implications of such a steady-state equilibrium model for the composition of the source region for arc magmas are that the more fluid compatible elements which are able to move across the flow will tend to maintain their original fluid concentrations which, if adequately enriched, will lead to enrichment of the matrix in the source region. In contrast, the concentration of elements with higher partition coefficients which are carried down with the wedge will be unchanged in the matrix of the source region.

The degree to which the fluid will interact with the matrix and modify its minor and trace element abundances depends on the relative volumes of fluid required to achieve equilibrium with the matrix. To estimate the final source concentration after enrichment we can use limiting values of the original source and fluid concentrations. Thus the mass balance for the source region for the fluid dominated elements is

$$C_s = C_f^0((1-\phi)\,D_{br/f}+\phi), \tag{5a}$$

where C_f^0 is the initial fluid concentration, $D_{br/f}$ is the bulk rock/fluid partition coefficient in the wedge and ϕ is the mass fluid fraction present, while the mass balance for the matrix dominated elements in the mantle wedge source region is

$$C_s = C_w^0((1-\phi)+\phi/D_{br/f}), \tag{5b}$$

where C_w^0 is the initial source or matrix concentration. If the fluid was in equilibrium with the slab then (5*a*) can be expressed in terms of the slab matrix concentration as

$$C_s = (C_{sl}^0/D_{sl/f})\,((1-\phi)\,D_{br/f}+\phi). \tag{5c}$$

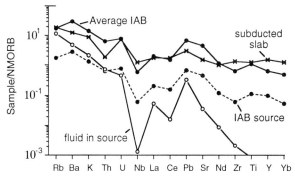

Figure 5. Minor and trace element contents of average island arc basalt; the subducted slab (both from Hawkesworth *et al* 1991*a*); an island arc basalt source calculated assuming for the sake of discussion that average arc basalt is generated by 10% fractional melting of spinel peridotite; and the fluid in equilibrium with that arc basalt source calculated by using the rock/fluid distribution coefficients in figure 4*a*. In the model discussed here, the source for average island arc basalt is developed from a depleted mantle composition (NMORB/10) by interaction with fluid of the composition shown.

The minor and trace element abundances in an arc source were estimated assuming that an average arc basalt (Hawkesworth *et al.* 1991*a*) is the product of 10% fractional melting of spinel peridotite. Using the estimated rock/fluid D values the composition of the fluid in equilibrium with the arc source peridotite was then calculated, assuming 0.5% fluid by volume, although these calculations are not sensitive to the actual value of porosity provided it is small. The results of these calculations are illustrated in figure 5, and although the fluid component is negligible in terms of mass balance within the source region, because $D_{\mathrm{br/f}} > \phi$ and $\phi \ll 1$, the fluid can clearly have a significant effect in terms of modifying the matrix composition, provided local equilibrium is maintained.

To illustrate how a source region may develop over time in the mantle wedge, we can use equation (2) to estimate the minimum velocity required for individual elements to move across the wedge and by specifying a distance to the source region, we can calculate the travel times for each element. For a slab dipping at 45° and a wedge velocity of 30 km Ma^{-1} (Peacock 1991; Davies & Bickle 1991), then the fluid velocity must be greater than 21 km Ma^{-1} for even completely incompatible elements not to be dragged down with the wedge. For a distance of 10 km between the slab and the source region, the maximum travel time for the fluid is 0.47 Ma. However, because of the high D values for the fluid, the fluid velocities need to be much higher than the minimum, implying shorter travel times. In figure 6 we show the travel time as a function of fluid velocity for selected elements for a 10 km distance, U_{s} of 30 km Ma^{-1} and α of 45°. As expected from the relative partition coefficients, the more incompatible elements will be the first to reach the source region for a given fluid velocity. However, the fluid velocities required to enrich even the most incompatible elements are very high. Davies & Bickle (1991) estimated the mass flux of water from the slab to be *ca.* 2×10^5 kg m^{-2} Ma^{-1}, equivalent to an effective velocity of 40 km Ma^{-1} for 0.5% porosity. Although the flux estimates by Peacock (1990) are an order of magnitude higher than that of Davies & Bickle (1991), it is clear that the fluid velocities are unlikely to be as high as the results outlined in figure 6 suggest. The timescales, or velocities, could be reduced if the transport distance is significantly less than the 10 km specified here or, more likely, the estimated

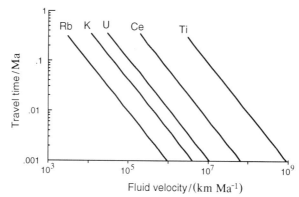

Figure 6. The time taken for the concentration of five elements in the fluid at a point 10 km from the fluid input zone to reach 90 % of the initial fluid values, for a range of fluid velocities. The initial matrix has the minor and trace element content of depleted mantle, and the initial fluid is that calculated to be in equilibrium with the average island arc basalt source in figure 5. If the fluid is in equilibrium with the matrix, the calculated timescales are equivalent to those required for the development of an arc signature in the source peridodite.

partition coefficients are unrealistically large. An order of magnitude reduction in the chosen partition coefficients will result in a broadly similar reduction in element transport times or velocities.

4. Discussion

A model in which the mantle wedge is modified by interaction with a fluid derived from the subducted slab can qualitatively explain many of the distinctive minor and trace element features of destructive plate margin magmas. Such differences between MORB and OIB on the one hand, and IAB on the other, are most readily explained by differences in the partition coefficients into silicate melts and fluids (Pearce 1982) rather than by differences in the residual minerals. Moreover, the available data are consistent with a systematic variation in rock/fluid distribution coefficients and ionic radii, at least in the range 69–167×10^{-12} m. Such fluid percolation models account for the key observation that even though island arc rocks are relatively enriched in LILE, the LILE abundances vary less than the HFSE, since it is the LILE which are more likely to reach equilibrium values. However, there are serious problems both in the inferred distribution coefficients outside the range 69–167×10^{-12} m and in the high values of the estimated rock/fluid distribution coefficients, which imply un-realistically high fluid velocities (figure 6).

B and Be have very small ionic radii (15×10^{-12} m and 30×10^{-12} m) and so using the extrapolation in figure 3, their inferred olivine/fluid partition coefficients are extremely high (greater than 10^5). However, island arc rocks are characterized by relatively high B/Be and B/LREE ratios, and in some areas by elevated ^{10}Be (Morris *et al.* 1990). Thus it has been argued that B is preferentially mobilized in slab derived fluids, despite its much smaller ionic size than for example the LREE, and that some Be in certain island arc suites is derived from subducted sediment (Morris *et al.* 1990). Moreover, the presence of ^{10}Be in some island arc rocks implies that the transfer time of Be from subducted sediment to the zone of magma generation in the mantle wedge is less than 7.5 Ma (five half-lives of ^{10}Be), and at least for the equilibrium model presented in figure 6 this would require fluid fluxes 3–4 orders of magnitude higher than those suggested by Peacock (1990) and Davies & Bickle (1991). Clearly the distribution coefficients for B and Be extrapolated on the basis of ionic radius are too high, either because in practice their mobility is dramatically increased by the development of complexes (e.g. borates), or rock/fluid D values decrease at low as well as high values of ionic radius and so the curve in figure 3 should not be extrapolated outside the range 69–167×10^{-12} m.

In addition to the question of whether the relative partition coefficients for B and Be are too high, it also appears that all the absolute rock/fluid partition coefficients are too high for realistic fluid velocities and time scales (figure 6). In principle radiogenic isotopes can be used as tracers to detect contributions from subducted material in arc rocks, although in practice this signature may be diluted by interaction with the matrix as the fluid moves towards the magma source region (figure 2). Hawkesworth *et. al.* (1991*a*) estimated maximum values for the slab derived contribution in an average arc basalt, assuming no isotope exchange with the mantle wedge. Clearly if exchange occurs such values may be significantly increased. None the less, using their value of 20% for the proportion of Sr derived from the subduced slab, the rock/fluid distribution coefficient would need to be 2–3 orders of magnitude less than that inferred from the Brenan & Watson results, assuming that the transport distance was 10 km and the timescale *ca.* 1 Ma. As discussed by Brenan & Watson (1991) the most likely cause of this discrepancy between the available experimental data and the evidence for element mobility from both high pressure eclogites (Philippot & Selverstone 1991) and mantle xenoliths (Vidal *et al.* 1989), is that the natural systems exhibit more extreme fluid compositions.

Additional constraints on the timescales of trace element transport from the subducted slab to the zone of magma generation are available from U-series disequilibria studies. The majority of subduction-related rocks show relatively little $(^{238}\mathrm{U}/^{230}\mathrm{Th})$ disequilibrium, but those that do tend to have high $(^{238}\mathrm{U}/^{230}\mathrm{Th})$, and occur in depleted island-arc suites (Gill & Williams 1990; McDermott & Hawkesworth 1991). The presence of excess U requires that U was enriched relative to Th over timescales of less than 350000 years before eruption, and this has been ascribed to U-enriched hydrous fluids derived from the subducted slab (Allègre & Condomines 1982; Condomines *et al.* 1988). If this explanation is correct, and if the disequilibrium is established primarily during slab-dehydration, then for the model discussed here the peridotite/fluid distribution coefficient for U would be not more than 0.001, also three orders of magnitude lower than the value in figure 4*a*. Some of the discrepancy may be because U mobility is sensitive to oxidation state as well as ionic radius, and some may be explained if much of the U excess is established during horizontal fluid transport within the mantle wedge rather than at the slab–wedge interface. None the less the fact that similar discrepancies are inferred for Sr and U reinforces the suggestion that mineral/fluid distribution coefficients must be lower than those used here, presumably because of differences in fluid composition. Additionally, however, we stress that these conclusions are to some extent model dependent in that local equilibrium has been assumed throughout.

Finally, there is the question of partial melting, and how that may vary in different island arcs. It is inferred that partial melting takes place beyond the zone of fluid percolation, as illustrated in figure 2, and several authors have argued that there is little evidence that partial melting beneath arcs results in different element fractionation trends than partial melting elsewhere (Plank & Langmuir 1988). Element ratios such as Ce/Sm may be used as indices of the degree of melting, assuming that some arc magmas are derived from source rocks with similar REE ratios, whereas LIL/HFSE and LIL/REE ratios may reflect fluid fluxes and the amount of water which reaches the magma source regions. The low Ce/Yb arc rocks have similar radiogenic isotope ratios and so they are most likely to have been derived from sources with similar trace element ratios. Furthermore, the systematic variation between average Ce/Sm, but not K/Sm, with crustal thickness in the low Ce/Yb suites (figure 1) suggests that whereas the degree of melting may vary with crustal thickness, this does not appear to be linked in any simple way to differences in the fluid contribution.

Both Plank & Langmuir (1988) and Davies & Bickle (1991) concluded that the degree of partial melting in arcs varies with crustal thickness, and hence with the length of the melting column. However, the inferred amounts of melting varied from 10–25 % to 2–8 % in the two studies, and it was clear that a correlation between the degree of melting and crustal thickness could result from adiabatic decompression and/or be a function of the fact that the mantle is colder in areas of thicker crust. Given that the degree of melting is a nonlinear function of the length of the melting column, and that Ce/Sm varies nonlinearly with degree of melting, the weak correlation between Ce/Sm and crustal thickness in figure 1 may be significant. However, until better data-sets are available we simply conclude that the observed range in Ce/Sm in the low Ce/Yb arcs is consistent with 3–18 % melting of slightly LREE depleted source rocks (see §2), which in the model developed for major elements presented by Davies & Bickle (1991) would suggest that the length of the melting column varies from *ca.* 15 km to over 50 km beneath different arcs.

We thank Huw Davies and Matthew Thirlwall for their thorough and constructive reviews.

References

Allègre, C. J. & Condomines, M. 1982 Basalt genesis and mantle structure studied through Th-isotope geochemistry. *Nature, Lond.* **229**, 21–24.

Arculus, R. J. & Powell, R. 1986 Source component mixing in the regions of arc magma generation. *J. geophys. Res.* **91**, B6, 109–133.

Brenan, J. M. & Watson, E. B. 1991 Partitioning of trace elements between olivine and aqueous fluids at high P–T conditions: implications for the effect of fluid composition on trace element transport. *Earth planet. Sci. Lett.* **107**, 672–688.

Condomines, M., Hemond, Ch. & Allègre, C. J. 1988 U–Th–Ra radioactive disequilibria and magmatic processes. *Earth planet. Sci. Lett.* **90**, 243–262.

Coulon, C. & Thorpe, R. S. 1981 Role of continental crust in petrogenesis of orogenic volcanic associations. *Tectonophys.* **77**, 79–93.

Davies, J. H. & Bickle, M. J. 1991 A physical model for the volume and composition of melt produced by hydrous fluxing above subduction zones. *Phil. Trans. R. Soc. Lond.* A **335**, 355–364.

Eggler, D. H. 1987 Solubility of major and trace elements in mantle metasomatic fluids: experimental constraints. In *Mantle metasomatism* (ed. M. A. Menzies & C. J. Hawkesworth), pp. 21–42. London: Academic Press.

Gill, J. B. & Williams, R. W. 1990 Th isotope and U-series studies of subduction-related volcanic rocks. *Geochim. cosmochim. Acta* **54**, 1427–1442.

Hawkesworth, C. J., Hergt, J. M., Ellam, R. M. & McDermott, F. 1991*a* Element fluxes associated with subduction related magmatism. *Phil. Trans. R. Soc. Lond.* A **335**, 393–405.

Hawkesworth, C. J., Hergt, J. M., McDermott, F. & Ellam, R. M. 1991*b* Destructive margin magmatism and the contributions from the mantle wedge and subducted crust. *Aust. J. Earth Sci.* **38**, 577–594.

Henderson, P. 1982 *Inorganic geochemistry*. (353 pages.) Pergamon Press.

Kay, R. W., Sun, S. S. & Lee-Hu, C. N. 1978 Pb and Sr isotopes in volcanic rocks from Aleutian Islands and Pribilof Islands, Alaska. *Geochim. cosmochim. Acta* **42**, 263–273.

Leeman, W. P. 1983 The influence of crustal structure on compositions of subducted-related magmas. *J. Volcanol. geotherm. Res.* **18**, 561–588.

McDermott, F. & Hawkesworth, C. J. 1991 Th, Pb, and Sr isotope variations in young island arc volcanics and oceanic sediments. *Earth planet. Sci. Lett.* **104**, 1–15.

McKenzie, D. 1984 The generation and compaction of partially molten rock. *J. Petrol.* **25**, 713–765.

McKenzie, D. & O'Nions, R. K. 1991 Partial melt distributions from inversion of rare earth element concentrations. *J. Petrol.* **32**, 1021–1091.

Morris, J. D. & Hart, S. R. 1983 Isotopic and incompatible trace element constraints on the genesis of island arc volcanics from Cold Bay and Amak Island, Aleutians, and implications for mantle structure. *Geochim. cosmochim. Acta* **47**, 2015–2030.

Morris, J. D., Leeman, W. P. & Tera, F. 1990 The subducted component in island arc lavas: constraints from Be isotopes and B–Be systematics. *Nature, Lond.* **344**, 31–36.

Navon, O. & Stolper, E. 1987 Geochemical consequences of melt percolation: the upper mantle as a chromatographic column. *J. Geol.* **95**, 285–307.

Peacock, S. M. 1990 Numerical simulation of metamorphic pressure–temperature–time paths and fluid production in subducting slabs. *Tectonics* **9**, 1197–1211.

Pearce, J. A. 1982 Trace element characteristics of lavas from destructive plate boundaries. In *Andesites: orogenic andesites and related rocks* (ed. R. S. Thorpe), pp. 524–548. New York: Wiley.

Pearce, J. A. 1983 Role of sub-continental lithosphere in magma genesis at active continental margins. In *Continental basalts and mangle xenoliths* (ed. C. J. Hawkesworth & M. J. Norry), pp. 230–249. Nantwich: Shiva.

Philippot, P. & Selverstone, J. E. 1991 Trace-element-rich brines in eclogite veins: implications for fluid composition and transport during subduction. *Contrib. Mineral. Metrol.* **106**, 417–430.

Plank, T. & Langmuir, C. H. 1988 An evaluation of the global variations in the major element chemistry of arc basalts. *Earth planet. Sci. Lett.* **90**, 349–370.

Tatsumi, Y., Hamilton, D. L. & Nesbitt, R. W. 1986 Chemical characteristics of a fluid phase released from a subducted lithosphere and origin of arc magmas: evidence from high pressure experiments and natural rocks. *J. Volcanol. geotherm. Res.* **29**, 293–309.

Tera, F., Brown, L., Morris, J., Sacks, I. S., Klein, J. & Middleton, R. 1986 Sediment incorporation in island-arc magmas: inferences from [10]Be. *Geochim. cosmochim. Acta* **50**, 535–550.

Vidal, P., Dupuy, C., Maury, R. & Richard, M. 1989 Mantle metasomatism above subduction zones: trace-element and radiogenic isotope characteristics of peridotite xenoliths from Batan Island (Philippines). *Geology* **17**, 1115–1118.

White, W. M. & Dupré, B. 1986 Sediment subduction and magma genesis in the Lesser Antilles: isotopic and trace element constraints. *J. geophys. Res.* **91**, B6, 5927–5941.

Woodhead, J. D. & Fraser, D. G. 1985 Pb, Sr and [10]Be isotopic studies of volcanic rocks from the northern Mariana Island. Implications for magma genesis and crustal recycling in the Western Pacific. *Geochim. cosmochim. Acta* **49**, 1925–1930.

Discussions

Chapter 1

R. Thompson (Durham)   I have two questions. The first is whether you have compared ion probe data for the minerals of ocean island basalts and lamproites? The second concerns whether the rocks you have been studying have been metamorphosed. They have been at higher pressures and temperatures for longer periods than have Lewisian gneiss for example. In the Lewisian you often find a boudin of serpentinized peridotite, surrounded by acid gneiss with which it reacted strongly to produce a whole series of monominerallic zones. Why then are the rocks you have been studying to be interpreted as igneous rather than as metamorphic rocks, with veins of magma intruding such metamorphic zones?

Harte   In reply to your first question, we have compared calculated trace element data for metasomatic fluids (I would say melts) with trace element data for kimberlites and lamproites. Data for the potential metasomatic melts in equilibrium with amphibole-bearing (for example, potassic richterite-bearing) metasomatic xenoliths from kimberlites, have been determined from analyses of minerals in xenoliths coupled with application of appropriate partition coefficients to calculate the melt compositions. The compositions of the calculated metasomatic melts/fluids (for example, Figure 6a of our chapter) overlap those of kimberlites and lamproites, at least for REE. Menzies *et al.* (1987) (*Mantle metasomatism* (ed. M.A. Menzies and C.J. Hawkesworth). Academic Press, London; Figure 4b, p. 327) have shown such close similarities, though they argue for a hydrous fluid as the cause of metasomatism. We are suggesting that the metasomatic agent giving rise to minerals like potassic richterite, is an evolved melt, and that the trace element charateristics of such metasomatic melts and those of kimberlites and lamproites may arise by processes of crystal fractionation as the melts migrate through, and maintain contact with, mantle peridotite. Earlier stages in that differentiation process, from an initial melt with OIB-like trace element characteristics, may be represented by Matsoku-type metasomatic rocks, where the melt was precipitating considerable garnet and ilmenite in addition to pyroxenes and olivine.

 Concerning your second question, one generally does not see evidence of discrete metamorphic events causing changes as one does in the crust, and it is more of a question of subsolidus re-equilibration during cooling. It is common for the peridotite wallrocks adjacent to mantle dykes and veins to show modified compositions, whereby the wallrock minerals show compositions like those inside dykes and veins; however the magnitude of such changes is relatively small (for example a few percent in $Mg/(Mg+Fe)$) and rock types as disparate as acid gneiss and peridotite are not seen in contact. On the whole, a number of factors, such as width of the mantle wallrock reaction zones, suggest that they largely develop during times when melt is present, because diffusivities of elements in minerals are too small. The extent to which trace element mineral compositions are modified by subsolidus metamorphic equilibration, is hard to judge. Evidence is provided by the array of compositions seen for example in clinopyroxenes; if one takes megacrysts (phenocrysts?) or crystals from metasomatized peridotites, one

Phil. Trans. R. Soc. Lond. A (1993) **11**, 193–215

Printed in Great Britain

finds a wide array of overlapping compositions, and one might not expect such similar overlapping ranges if the clinopyroxenes from metasomatized xenoliths had undergone extensive subsolidus metamorphic reconstitution.

Frey I thought that kimberlite magmas are rich in nickel. If so it would be difficult to derive them by fractionation from an ocean island basalt.

Harte Most kimberlite magmas certainly are rich in nickel, though it is not simple to separate the nickel introduced by disaggregation of the wall rock from that in the original melt. It is also unclear to what extent the fractionating melts remain in chemical equilibrium with the peridotite matrix through which they are injecting and percolating. If they do remain in local equilibrium, their nickel content will be buffered by that of the peridotite. Please note, however, that we do not wish to produce kimberlite by simple fractional crystallization of ocean island basalt. Our initial magma would be more ultrabasic, but it would have OIB-like trace element distributions. I believe that fractional crystallization is important because it can reproduce the trace element variations seen in kimberlite magmas, even though such magmas may often freeze in the lithosphere before subsequently being melted and erupted.

Cox In some metasomatized nodules, particular minerals have been produced by reaction at grain boundaries. For instance, you commonly find that phlogopite has been produced on the boundaries of garnet. I have always thought that this behaviour was caused by percolating fluid that was deficient in alumina, and that it reacted with the garnet. I would not refer to such a fluid as a magma. You have argued that all so-called metasomatism results from reaction with magma, yet you are still using the term 'metasomatism' to describe what is going on. I find this nomenclature confusing.

Harte An early definition of metasomatism goes back to Goldshmidt (1922) (*Economic Geology*, **17**, pp. 105–23), who defined it in terms of changing the composition of an essentially solid rock. He did not specify whether this change occurred through a reaction with a volatile-rich phase or a melt, but clearly accepted both to be possible. I think that many people have assumed that the metasomatic agent was a volatile-rich fluid because of an error in understanding the petrographic evidence. I think that the principal reason why fluids, rather than melts, were thought to be involved was because melts were not supposed to be able to move through a crystal matrix until the melt fraction was about, or perhaps exceeded 10 per cent. Such a melt fraction would be easily seen petrographically. Once it became clear that this view was not correct, and that melt fractions as small as 1 per cent or 0.1 per cent could move (for example, McKenzie (1985). *Earth Planet. Sci. Lett.*, **74**, 81–91), melt movement could produce the effects that had previously been attributed to volatile-rich fluids. Thus, the principal reason why volatile-rich fluids were so widely believed to be responsible for metasomatism was because of this mistaken view about melt movement involving such large volumes of material that it was bound to be evident petrographically. I think that the movement of small melt fractions can also account for most of the small-scale effects previously assigned to granitization, and there is then no need to appeal to nebulous hypotheses that no one understands.

  With respect to your comments concerning phlogopite and garnets, some of the reactions that produce phlogopite rims round the garnets occur as the rocks move towards the surface. The kimberlitic magma penetrates the rock along the grain edges, and causes the incipient alteration that is so commonly observed. The melt may tend to pool around the garnets because they are more resistant to deformation than is the olivine matrix.

Cox I was really more concerned with the thicker replacement rims that occur in the coarse-grained nodules, where you see the sequence of garnet with phlogo-pite rims, to phlogopite with small amounts residual garnet, finally changing to phogopite and potassic richterite. Surely these reactions do not occur during the upward transport by the kimberlite?

Harte Yes, I agree that those reactions are not simply a consequence of de-compression and eruption. However, I would suggest that the evolving (partly by crystal fractionation) metasomatic melt reaches a reaction point with garnet under some conditions and this leads to the textures you note. I do not see why this should not happen with melt as well as with volatile-rich fluid.

Kurz You implied that the metasomatic reactions which you discussed occurred entirely within the lithosphere. Do you believe that the same processes also occur in the asthenosphere?

Harte I emphasized the lithosphere because all the nodules that I discussed were clearly derived from the lithosphere. We have few, if any, nodules from the asthenosphere. The only nodules that may come from the asthenosphere are peridotites and are highly deformed. They are full of small olivine neoblasts. Such nodules are commonly referred to as sheared nodules. These nodules form at very high temperature and have certainly undergone some metasomatism. Changes in the iron–magnesium ratio and in the titanium content are similar to those that occur in higher-temperature lithospheric wallrock xenoliths. But the sheared nodules do not show any evidence of modal metasomatism, even though melt appears to have passed through them.

Chapter 2

McKenzie What you have talked about is, I think, the simplest possible case. The real geological problem is likely to be more complicated, because the equations you have used do not describe the flow when the melt has segregated into channels such as dykes. An approximate method of including such effects would be to allow the grain size to vary, but there is no expression in your equations that relates the grain size to the other physical parameters. In your calculations the grain size is kept constant. Any attempt to take such behaviour into account must make the problem more complicated.

Spiegelman Yes, these equations must break down when the melt moves through the lithosphere, partly because the matrix is in the brittle regime and does not behave like a viscous fluid, and partly because the Reynolds and the thermal Peclet numbers of the melt are no longer very small. Otherwise the matrix would solidify before it could be erupted. But beneath ridges the upwelling rate of the matrix is likely to be less than about 10 mm a^{-1}, so the melting rates will be slow. The melt channels provide a natural mechanism for producing local regions of high porosity, and hence for producing disaggregation. Once that happens the equations are no longer valid.

White   The location of the freezing zone in your models is fixed spatially, so if the freezing rate is large, the porosity in this zone will quickly reduce to zero. Then, the melt will accumulate below this zone and produce a melt channel. In the real Earth the freezing zone will not be fixed, and may be able to move downwards to solidify all the melt that is rising through the crust. Could the production of melt channels in your models be a result of forcing the location of the freezing zone to be fixed?

Spiegelman   I believe that the models I have used are relevant to melting beneath ridges, because the production of channels depends on the local structure of the freezing zone. The geometry of the models is not meant to represent a ridge. The two important features that produce the channels are that the lower boundary of the freezing region is inclined, and that the melt is removed by freezing over a distance that is comparable to or smaller than the compaction length. The channel geometry will clearly vary spatially beneath a spreading ridge, but will always be controlled by the local freezing rate.

McKenzie   It is important to emphasize that the freezing rate is imposed on all the models that you have discussed. In fact the freezing rate must be thermodynamically controlled, by the energy equation, and could be quite different in its spatial distribution from that in any of the models you have discussed.

Spiegelman   What you say is correct. But simple estimates of the heat transported by 1 per cent melt percolating through the matrix suggest that advection of heat by the melt can be neglected. So I think that the conductive cooling model is actually quite a good approximation.

M. Kurz   How do the models you have discussed produce spatial and temporal variations in the geochemical composition of the melt? Based on your calculations, on what scales do you expect these variations to occur?

Spiegelman   Each element requires the solution of one additional second-order differential equation. If solitary waves are present in the source region, their arrival at the surface will produce a rapid increase in the melt flux. The solitary waves are waves of porosity and travel faster than the melt itself, and therefore faster than any geochemical signal, even of the most incompatible element. So any geochemical signal that is initially contained in a solitary wave is left behind as the wave moves upwards. Sarah Watson has done a number of numerical experiments to illustrate how chemistry and porosity become decoupled in these systems. The passage of a solitary wave through a chemically enriched layer disperses the enrichment over a wider zone, so the presence of solitary waves in the melting region can have an important influence on the trace element distribution.

Singh   The amplitude of the one-dimensional solitary waves in your experiments is greater than that of those in the two-dimensional models. Is this difference a numerical artifact?

Spiegelman   I think that the numerical artifacts in these experiments are small. Some of the waves are of very similar amplitude in one and two dimensions. Furthermore, in some of the two-dimensional calculations the matrix is also allowed to move. The existence of the channels is quite strongly affected by such advection, but I need to do more experiments in both one and two dimensions before

I fully understand the behaviour. One of the difficulties is that the equations are strongly nonlinear.

McKenzie You remarked that the presence of solitary waves might lead to a large-scale anisotropy in the melting region. I would be surprised if this were correct, because I believe that two-dimensional solitary waves will be unstable to three-dimensional disturbances, in the same way as one-dimensional disturbances. If this argument is correct, melt movement should occur by the movement of spherical solitary waves that are not likely to produce any large-scale anisotropy. Only near your freezing zone will the boundary conditions stabilize the two-dimensional waves.

Spiegelman I agree with this. Indeed the generation of two-dimensional waves can clearly be seen in Fig. 6. In this case the two-dimensional disturbance is produced by reflection from the left-hand boundary of the region, and is therefore an artifact of the calculation. But similar effects should occur beneath a ridge axis and in the head of a plume, where waves can propagate from one side of the region to the other. So such instabilities are likely to exist everywhere.

McKenzie If such three-dimensional solitary waves are present, it is no longer obvious that the thermal Peclet number will be small. There is then the possibility that other instabilities can arise through coupling between melt movement and the freezing rate, and controlled by the energy, rather than the momentum, equation.

Spiegelman Such behaviour may well occur, but I have not yet attempted to explore such complicated systems.

Frey Would you expect the passage of solitary waves to leave a geochemical signature in the residual peridotites?

Spiegelman This is a hard question to answer. I do not yet know whether or not there will be any measurable effects left after the passage of many waves through the same region. I don't think we have ever seen a convincing example of a solitary wave frozen into a rock. The thickness of the channel is likely to be about a kilometre, and it will not be easy to recognize pieces of such a channel brought up as nodules or as hand specimens.

Furthermore, the porosity in the channel is about 4 per cent, which is not very different from that of 1 per cent in the rest of the region. Such a variation will not be easy to recognize when the melt has solidified, though it might be possible to do so by geochemical means.

Chapter 3

McKenzie Your theory assumes that the grain size is constant everywhere. How good an assumption is this?

Kohlstedt The grain size actually varies rather little in each experiment. A worse problem is indicated by the faceted faces on many of the grains in the photomicrographs. These show that the melt geometry differs from that which we assume to calculate the permeability and the surface energy. Another limitation may be the assumption that the matrix deformation is viscous and is rate-limited by diffusion. But the agreement between the values we obtain for the matrix viscosity and those measured by Karato (Karato *et al.* (1986). *J. Geophys. Res.*, **91**, 8151–76) suggests that this assumption is valid.

McKenzie   Your images show a variation in backscattered intensity in the olivine close to the melt channels. Is it caused by variations in the magnesium number? If so there must be an exchange reaction occurring between the melt and matrix, and this could have an important effect on melt migration at these extremely small melt fractions.

Kohlstedt   These systems are close to chemical equilibrium, though small variations in the iron–magnesium ratio are present. But the variations in backscattered intensity are principally caused by topography rather than by compositional variations. To see the melt geometry clearly we have to etch the surface. The resulting topography produces artifacts due to double scattering. It is these that you notice in the micrographs and not the compositional variations.

McKenzie   I do not understand how you calculate the melt percolation velocity into a region where the melt fraction is initially zero. There is a singularity in the equations, and the time taken for melt to penetrate into an unmelted region must always be infinite.

Kohlstedt   We avoid this problem by starting the calculation with a small melt fraction of 10^{-2} or 10^{-3} per cent in the unmelted region.

McKenzie   But the result you then obtain from the calculations will depend entirely on the initial value you choose.

Kohlstedt   Yes, but it makes very little difference whether you use 10^{-2} or 10^{-3} per cent.

McKenzie   This lack of influence must be due entirely to errors in the numerical scheme you use at small melt fractions. There is a real singularity at porosity, and an accurate numerical scheme should reproduce it.

Kohlstedt   Yes.

McKenzie   This difference with the calculations makes me concerned about whether the permeability does indeed depend linearly on the porosity, as you have argued. As Frank argued more than twenty years ago (Frank, F.C. (1968). Two components flow model for convection in the Earth's upper mantle. *Nature*, **220**, 350–2), at small melt fractions the permeability should vary as the square of the porosity.

Frank   I wish you had all got going on these problems twenty-five years ago!

Kohlstedt   I am also a little concerned about the validity of the linear dependence of permeability on porosity. But experimentally there is no doubt that the initial migration of the melt is very fast. It is this behaviour that requires the linear dependence. This behaviour has important implications for melt migration and for the mechanical properties of the partially molten system.

Osmaston   The material that is generating the melt must be able to flow into the relevant region while it is solid. The viscosity of the solid material is likely to be much greater than that of the partial melt. The effective viscosity of the material will therefore be a strong function of temperature. Such viscosity variations are likely to produce instabilities in the form of diapirs, especially beneath mid-ocean ridges.

Kohlstedt I am not convinced that the small amount of melt that is likely to be present will have a large effect on the rheology. A number of experiments have now been carried out with 5–10 per cent melt and show that such melt fractions do not have a large effect on the rheology. If the melt fraction that is present in the source regions beneath ridges is less than 5 per cent, as a number of people now believe, it probably has little effect on the rheology of such regions. Only if the melt geometry changes in such a way as to allow the grains to disaggregate will melt fractions of less than 10 per cent strongly affect the rheology. If the melt and matrix are in textural equilibrium, 10 per cent melt is too small a melt fraction for disaggregation to occur, and the interconnected grains largely carry the applied load. The melt increases the rate at which diffusion can occur, but does not allow the grains to slide past each other. If, however, the stresses are sufficiently large to cause microfracturing, the surfaces can be lubricated by the melt, which will then lead to enormous changes in the rheology. But we believe that the stresses in partially molten regions within the Earth are not sufficient to cause such behaviour.

Chapter 4

Takahashi Can your model explain why komatiites were formed in the Archaean, but not at later times?

Ceuleneer In our model the presence of melt acts to average the temperature distribution throughout the region that is melting at the head of the plume. The resulting temperature will be close to the average mantle potential temperature, whatever the temperature contrast was before the initiation of melting. In the Archaean this mean temperature was probably about $300\,^{\circ}\mathrm{C}$ greater than it is now. Both Archaean ridges and hotspots would therefore produce komatiites in the same way as they are now producing basalts.

R. White It is well known that the MgO concentration of tholeiitic basalts found on mid-ocean ridges, in flood basalts, and on ocean islands is always about the same. The explanation following Huppert, Sparks, Stolper, and Walker is that fractionation occurs near the base of the crust, rather than in the convecting region, and that only the fractionated basalts with their reduced density erupt through the crust. I have not understood why you require fractionation to occur in the convecting part of your model.

Ceuleneer The magnesium concentration can be produced by removal of olivine at high or at low pressure. I think that their proposal, that MgO-rich melts do not erupt at the surface because of their high density, is likely to be correct. But a number of other geochemical indicators, like the depletion in HREE, show that ocean island basalts are produced in a region that contains garnet. Shallow differentiation alone cannot explain the observed geochemical trends.

McKenzie Did you use two-dimensional axisymmetric or cartesian models for your comparison of the temperatures in the two- and three-dimensional models?

Ceuleneer We used two-dimensional cartesian calculations for this comparison. The axisymmetric calculations agree well with the three-dimensional ones, but we believe it is more satisfactory to allow the plumes to form as a stable solution to the full three-dimensional equations, rather than imposing axisymmetry on the solution.

Chapter 5

Kurz It is not clear to me whether you are arguing for a mantle that only contains two components, in which all the isotopic variability is produced by percolation processes? Even if such percolation processes are important, it is still very difficult to produce all the observed geochemical variations from a two-component mantle.

O'Nions I agree. But, as far as the rare earth elements are concerned, we have not yet been forced by observations to consider anything more complicated than a two-component mantle consisting of a primitive and a depleted component. But the behaviour of the very incompatible elements is undoubtedly more complex. Part of the complexity results from radioactive decay within the lithospheric boundary layer, and also perhaps deeper within the mantle. Whether you can identify the end members in magmas depends on whether the composition of the end members is conserved through the melt extraction process. I am not convinced that such conservation occurs. However, melts extracted from the well-mixed regions outside these boundary layers show essentially no isotopic or compositional variations. This uniformity is often forgotten because isotopic ratios can now be measured with an accuracy of a few ppm. But the variations observed within melts extracted from these regions are very small compared with those in melts extracted from the boundary layers. Since the residence time of some of the most incompatible elements in the convecting upper mantle is as short as 0.5 Ga, stirring in such regions must be very efficient. What we have done so far is to use this well-mixed and depleted region as one of our source compositions, but it is clear from both the rare earth element compositions and from the isotopic ratios that parts of the lithospheric mantle have been strongly enriched in incompatible trace elements, and that this enrichment occurred more than 1 Ga ago.

Wood How sensitive are your calculated trace element compositions to the melt distribution in the mantle? Have you carried out a sensitivity analysis, by forward modelling for instance? I believe that the observed concentrations are not very sensitive to the melt distribution. Figure 1 shows the best fit and uncertainties for the MORB rare earth element concentration ratios from McKenzie and O'Nions, calculated using the depleted upper mantle concentrations. The points are the observed ratios and the error bars show one standard deviation, obtained from the observations. I tested the sensitivity using simple forward modelling, and tried three arbitrary melt distributions. All generated no melt at depths greater than 80 km, the depth at which spinel is entirely replaced by garnet. The first is a step function, with a melt fraction of 7 per cent to a depth of 80 km. The second has a melt distribution that increases with increasing depth, from 3 per cent at the surface to 11 per cent at 80 km. The third is similar to an adiabatic melt distribution, though with a potential temperature less than 1280 °C. I did not attempt to change the melt distributions to fit the observed ratios. The calculated ratios from these three models all look very similar (Fig. 1a). That from McKenzie

Table 1.

	H	(obs.-calc.)/S.D.	rms(gradient)	rms(second derivative)	Crustal thickness (km)
Best fit	0.5113	0.1898	1.031×10^{-2}	6.771×10^{-5}	6.52
Near adiabatic	0.9262	0.4634	1.264×10^{-2}	1.092×10^{-4}	3.50
Step	0.8437	0.2161	1.941×10^{-2}	6.028×10^{-4}	7.60
Increasing	1.124	0.1953	3.051×10^{-2}	9.908×10^{-4}	7.99

The second column shows the value of H calculated from McKenzie and O'Nions (1991, eqn. 25).

and O'Nions (1991) has a small europium anomaly because melt is generated in the plagioclase peridotite stability field. It also has a slight heavy rare earth depletion because a little melt is generated in the garnet stability field. The ratios calculated from the step function are very similar to those from McKenzie and O'Nions' distribution, as are those from the model in which the melt fraction increases with increasing depth. Only the adiabatic model generates ratios that are appreciably different from those observed. These three melt distributions are very different from each other, and two are well outside the error bounds given by McKenzie and O'Nions. So I think the melt distributions that you obtained are controlled by the other constraints you imposed on the inversion.

O'Nions Only some of the melt distributions you show are compatible with our understanding of the physical processes that generate melt. As we made clear, our aim was to obtain melt distributions that were compatible with both the trace element concentrations and the physical constraints on melt generation. We think that it is unreasonable to allow the melt fraction to increase with depth in regions where the temperature structure is dominated by the upward advection of heat. This is the reason why we weighted the inversion strongly against melt fractions that increased with depth when we calculated the penalty function. As Table 1 shows, it is principally the gradient terms in the penalty function that differ between our melt model, the step model, and the one in which the melt fraction increases with depth. As you remark, these constraints are not imposed by the geochemical observations, but by our views about the physics of melt generation.

Wood But the important point is that the trace element ratios are not very sensitive to the melt distribution. Even the model in which the melt increases with depth fits the observations well, yet it is well outside your error bounds.

McKenzie It is important to be clear about what the rare earth element ratios can and cannot tell you. The first question concerns how many independent pieces of information are contained in the rare earth element ratios. This is an important concept in inverse theory. The ratios in Fig. 1a only contain two or three such pieces of information. Other distributions we used, especially those from the Hawaiian tholeiites, contain many more. But for the MORB data, we cannot hope to obtain more than two or three constraints on the melt distribution with depth from the rare earth element ratios alone. The most important constraint is on the total amount of melt that is made, and the reason why three of the four models fit the observations well is that they produce essentially identical amounts of melt (Table 1). Only the adiabatic model does not, and the difference between the observed and calculated ratios for this model is about twice that of the others (Table 1).

An important ambiguity in the melt distribution occurs if melt is generated in a region of constant mineralogy. For our purposes the plagioclase and spinel peridotite fields are essentially the same, and only melt generated in the garnet field has a significantly different signature. When the mineralogy is constant, the calculated ratios are unchanged if the melt distribution is inverted, so that it increases rather than decreases with depth. This symmetry is exact:the value of the relevant integral is invariant under such a transformation. The result depends on the partition coefficients being independent of pressure and temperature, as they are in both our and your calculations. The reason why we weighted positive and negative gradients of the melt distribution with depth differently was to remove this ambiguity. This is the reason why your model in which the melt fraction increases with depth also reproduces the observed ratios.

The reason why the step function fits is somewhat different. As Fig. 1b shows, its melt distribution lies almost within our error bounds. These bounds were calculated using an approximate expression, but, even when the most sophisticated inverse theory is used, the error bounds of the model are its least well-determined feature. So no importance should be attached to the fact that the step function does not everywhere lie quite within the error bounds: it is a perfectly satisfactory melt distribution from a geochemical point of view.

Our aim was to use our physical understanding of melt generation to resolve the inherent geochemical ambiguities. I do not believe that a melt fraction that increases with depth, at depths of more than about 10 km below an oceanic ridge, is compatible with melt generation by adiabatic decompression, and would be interested in your reasons if you think otherwise.

Wood   I agree with these arguments, but still think that you give unrealistically small error bounds for your models. The models that I used are not based on physical processes, but then neither are your melt distributions obtained by inversion.

O'Nions   The most important point about all the distributions that you show is that none of them involves the generation of appreciable amounts of melt in the garnet stability field. If one did, the calculated ratios would no longer fit those observed.

Wood   But the step model does generate a considerable amount of melt in the garnet stability field, below 60 km.

O'Nions   It is the field in which garnet alone is stable, below 80 km, that is critical to this argument.

McKenzie   The calculated ratios are controlled by the extraction efficiencies, which are illustrated for the four models in Fig. 2. These show that the Sm/Yb ratio is strongly affected by the presence of garnet, even when the melt fraction is as large as 11 per cent in the step model. There is no such difference in the observed ratios, and it is the absence of this effect that requires all models that

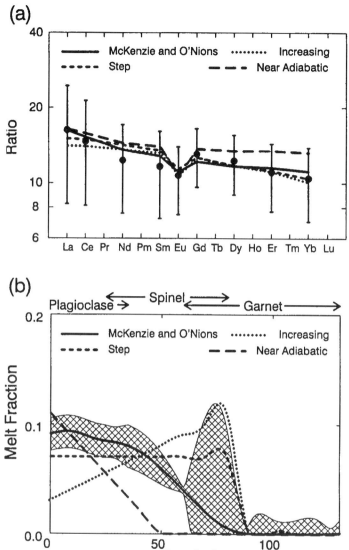

Figure 1. The lines in (a) show the rare earth element ratios with respect to depleted mantle, calculated for the four melt distributions in (b). The points in (a) show the observed ratios for basalts dredged from mid-oceanic ridges (McKenzie and O'Nions (1991), Fig. 3), and the error bars show one standard deviation. The shaded region in (b) shows McKenzie and O'Nions' estimate of the error envelope obtained from the inversion. The oscillations in the melt distributions in (b) for the step distribution, and for that which increases with depth, are caused by the influence of the discontinuity in the melt functions at 80 km on the cubic spline interpolation.

are to fit the data to generate the bulk of their melt in the spinel and plagioclase stability fields.

Chapter 7

McKenzie Though Keith and I used a discrete layer to discuss percolation effects (McKenzie and O'Nions 1991), we did so because the resulting behaviour

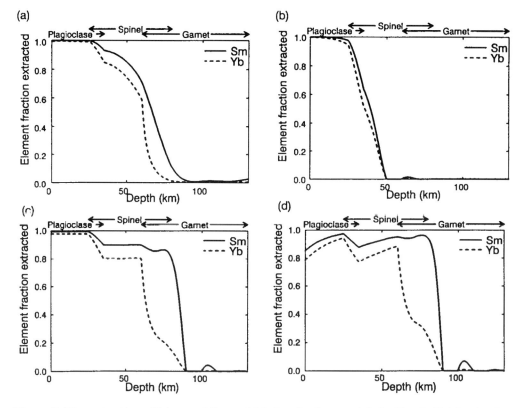

Figure 2. The extraction efficiency of Sm and Yb for the four models in Fig. 1. Melting in the field in which garnet is stable produces a large value of Sm/Yb that is not observed. Therefore, little of the melt can be produced in this region. (a) From McKenzie and O'Nions (1991); (b) near adiabatic model; (c) step model; (d) model with melt fraction increasing with depth.

could be described using simple analytic expressions. Similar effects occur when melt generation is continuous if the melt percolates through the asthenosphere that is being sheared, as it must beneath Hawaii. Then the angle between the vertical and the transport direction of any element will increase with the partition coefficient, and incompatible elements such as [3]He from the plume will reach the surface closer to the present location of the plume than will more compatible elements such as Sr. Percolation effects can therefore be important when melt generation is not episodic.

Kurz   But I do not understand how the observed differences in major and trace element composition between Kilauea and Mauna Loa can be produced by percolation effects alone. I believe that there is good evidence for heterogeneity within the plume itself, which is present before melting starts.

Thompson   I think that your results illustrate the difficulty of using geochemical observations to characterize the source regions of the magma. What at first appear to be unambiguous characteristics of the source region always turn out to be affected by processes that occur between the source region and the surface.

Some time ago, Wright and Fiske (1971) published a careful study of variations in petrography and chemistry of Hawaiian magmas over a twelve-year period (*Journal of Petrology*, **12**, 1–65). They thought that they could distinguish variations in the composition of small magma batches over periods as short as 10–100 years. I wondered if you had measured the He isotopic ratios on their suite of samples?

Kurz   We have been trying to do exactly what you suggest, but with a different suite of samples. In Hawaii we can look at variations that occur over a shorter time scale than is possible elsewhere. Rhodes and Frey have been measuring the major and trace element compositions of the same suite of rocks that I have been using for isotopic measurements. The trace elements do not show a simple variation with time, like the one I showed for He from Mauna Loa. The trace element behaviour therefore appears to be decoupled from that of the isotopes (Rhodes, personal communication). However, there are still few samples from any ocean island for which we have complete major element, trace element, and isotopic data, so it is still too early to tell.

Langmuir   We are all trying to construct physical models for melt generation and movement that can be tested. The percolation model that can account for the Sr and He data cannot also account for the behaviour of other incompatible elements, like Th or the light rare earths. Do you believe that you are justified in excluding some types of data when you are trying to find models that account for your observations, and if so why?

Kurz   I agree and think it is very important to use all possible types of field and geochemical data together, to try to understand the processes involved. But collecting such data sets is slow, and involves many people in different laboratories. At present I only have isotopic observations on these rocks. I am not intentionally excluding major and trace element data from the discussion, but there is just not yet enough overlap between the data sets. We are working on it.

Chapter 8

Osmaston   Can you obtain any information from komatiites about what the iron/magnesium ratio was in their source regions?

Takahashi   I have carried out the same sort of calculations for Archaean komatiites, and found that their compositions are compatible with garnet peridotites from South Africa being the residue after melt generation. The residue from komatiite melts must have an olivine composition close to Fo_{94}. The experimental results suggest that such a composition can be produced by melting a normal mantle peridotite with Fo_{89} if the degree of partial melting is large.

Osmaston   So a source with Fo_{85}–Fo_{87} is enriched in iron with respect to the source regions of komatiites?

Takahashi   Yes.

Frey   Many people in the past have proposed that the Hawaiian source region is enriched in iron, though I thought that this idea had finally been abandoned. A new problem is the recent discovery by Clague *et al.* (1991) of glasses with an MgO content of ~ 15 per cent and forsteritic olivine microphenocrysts, up to $Fo_{90.7}$. I suppose that such melts may be compatible with a fractional melting model, but highly forsteritic olivines are not unusual in Hawaiian shield lavas.

Takahashi I believe that such olivines come from the refractory part of the Pacific plate that overlies the plume. R. Helz (1987) (Diverse olivine types in lava of the 1954 eruption of Kilauea volcano and their bearing on eruption dynamics. In: *Volcanism in Hawaii*, US Prof. Paper No. 1250, pp. 651–722) has shown that Kilauean picrites contain three types of olivine phenocrysts. The highly magnesian olivines are large and contain kink bands, and have therefore been subjected to considerable stress. I believe that these crystals are derived from part of the Pacific plate that has disintegrated.

Wood I am concerned about your measurements of the distribution coefficient of iron. In your experiments you observed a correlation between the distribution coefficient and the total iron and magnesium content of the melt. How do you know that this correlation results from changes in the melt, rather than in the solid composition? The solid solution of fayalite in forsterite is known to be non-ideal. Therefore, as the composition of the olivine changes, so will the distribution coefficient. This effect will have a large influence on the composition of the melt. The correlation is then not with the total iron and magnesium in the melt, but with the ratio of iron to magnesium in the solid.

Takahashi In the case of olivine–liquid divalent cation partitioning, there are no large differences between K_d partition coefficients determined in Fe-rich systems (Takahashi, E. (1978). *Geochim. Cosmochim. Acta*, **42**, 1829–44) and Fe-free systems or Fe-poor natural rocks (Watson (1977); Hart and Davies (1978); Takahashi and Kushiro (1983)). I therefore believe that most of the variations in $K_d(Fe/Mg)$, $K_d(Ni/Mg)$, and $K_d(Mn/Mg)$ which I have compiled so far are due to the compositional variation of the melt, rather than of the olivine.

I would like to emphasize that the inversion calculation that I have presented here uses only well defined K_d values (with uncertainties of less than 10 per cent). On the other hand, most geochemical models of trace element behaviour during magma generation and solidification are based on simple Nernst-type partition coefficients D. Because the D values do not represent realistic chemical reactions between stoichiometric mineral solid solution and magma, their variations are theoretically unpredictable. I therefore encourage the use of K_d partition coefficients for mineral solid equilibria, in addition to that of olivine. I have shown examples of such calculations in Takahashi and Irvine (1981) (*Geochim. Cosmochim. Acta*, **45**, 1181–1185).

Also let me remind you that manganese plays a very important role in my analysis of olivine fractionation paths. Because there are as yet few accurate analyses of the manganese concentration in igneous rocks, rather little attention has yet been paid to this element. Calculations using manganese may be more satisfactory than those using iron, because the oxidation state of manganese is better known. I therefore recommend that igneous petrologists and geochemists in the audience include manganese analyses in their trace element tables. Because manganese is traditionally included in the major element list, its concentration is only given to two significant figures. It should always be given to three significant figures, because the first digit is always unity.

Chapter 9

Moorbath Could you explain how you believe the variations in isotopic ratio are produced? I thought that the French and Icelandic petrologists (Sigmarsson, O., Condomines, M., Grönvold, K., and Thordarson, T. (1991). Extreme magma homogeneity in the 1783–84 Lakagigar eruption: origin of a large volume of evolved basalt in Iceland. *Geophys. Res. Lett.*, **18**, 2229–32) found associations between isotopic ratios and trace elements in Iceland that are similar to those you have described. They have argued that these effects are produced by the mixing of melts from the plume with those from the Icelandic crust. The model is similar to the MASH model that has been used in the Andes. The observed variations can then be produced by mixing variable amounts of the two end members.

Frey I understand that there is a paper in press in which both the Icelanders and the French (Hemond, Arndt, Lichtenstein, Hofmann, Oskarsson, and Steinthorsson. *J. Geophys. Res.*, in press) have abandoned the model to which you refer. There are significant differences in the neodymium isotopic ratios between the alkalic and tholeiitic lavas in Iceland. These differences are too large to have been produced in 20 Ma, and therefore cannot be explained by ageing within the Icelandic crust.

I did not really discuss the origin of the isotopic heterogeneities in the Hawaiian shields. Many have proposed mixing of plume and MORB-related components. While this process works well for post-shield lavas, it does not adequately explain isotopic data for shield lavas. I think it is significant that all the isotopic ratios for Hawaiian lavas are between the MORB field and bulk earth estimates. Therefore, there is no good evidence for a very enriched component (high $^{87}Sr/^{86}Sr$ and low $^{143}Nd/^{144}Nd$) in the source region. Other than heterogeneity within the plume, I do not have a satisfactory explanation for the isotopic data, and I do not think anyone else does either.

Thompson I am interested in the thickness that you propose for the stem of the plumes. The models you discussed seem to have rather thinner stems than do most of those in the literature, and I wonder if they are as narrow as those of Loper (1991), (*Tectonophysics*, **187**, 373–84) and Loper and Stacey (1983), who argued that their thickness was only about 10 to 12 km? I wonder whether such an estimate is generally accepted, and whether such narrow plumes would be more prone to a varicose instability that could produce diapiric blobs?

McKenzie It is important to understand why axisymmetric plumes are so popular. The principal geophysical and petrological features of plumes depend on the temperature structure in the region 100–200 km beneath the plate. The temperature differences are 100–200 °C and the viscosity is low. The effective Rayleigh number is then large, and a large number of grid points are required to model the flow numerically. Until recently the required resolution could only be obtained with a two-dimensional numerical scheme. Watson and McKenzie (1981) used a two dimensional axisymmetric scheme with up to 112 points in the vertical to explore the behaviour up to a Rayleigh number of 2×10^7. As Ceuleneer showed earlier, three-dimensional calculations are now practicable, but those that he showed had a Rayleigh number of only 10^6. He used about 96 points vertically, and used a Cray. None of these calculations includes the buoyancy forces that arise from melting. These are likely to be important in the melting region. I do not believe that mantle plumes are axisymmetric, even when they are beneath a stationary plate. I expect that the plume heads contain fully three-dimensional flows, consisting either of radiating sheets or of time-dependent blobs that are advected by the larger scale flow. This complicated flow is likely to be strongly depth-dependent, because the buoyancy produced by melting can only extend as deep as the melting zone. At present it is not possible to carry out fully resolved calculations in three-dimensions, even on the largest machines.

If the flow is as complicated as these remarks suggest, it is scarcely surprising that existing models of plumes only account for a few of the features that Frey has described. Many of these occur over length scales of about 25 km, which imposes constraints on the processes involved. The circulation that is responsible must have a similar length scale. Such small-scale features are probably restricted to the low-viscosity zone between depths of about 100 and 200 km. The observations also suggest that the geometry of this circulation is quite long-lived.

The large scale circulation that a number of people have used to examine the overall thermal structure of plumes probably provides a good guide to their major features, but it is no surprise that it cannot account for many of the geological observations.

White   Another important property of the mantle is that its viscosity is strongly temperature-dependent. Convection in such material has narrower, rising and broader, sinking plumes than does that in constant viscosity fluids, and this behaviour is also likely to be important in the mantle.

Kurz   Your data showed that the major element compositions from historical Kilauea and Mauna Loa basalts have always been distinct. Can you exclude the possibility that the major elements have varied over longer time scales, and that the major elements of the basalts from Kilauea and Mauna Loa were similar at some time in the past?

Frey   The trace element abundance ratios are similar in the oldest and youngest lavas from Mauna Loa. The same is true of the lead isotope data. The major element data are harder to interpret, because silicon, iron, and calcium are mobile in the Hawaiian environment. Such alteration is clearly seen in the Ninole lavas (Lipman *et al.* (1990). *Bull. Volcanol.*, **53**, 1–19). Despite these problems, lavas from the older 2–3 Ma Koolau Volcano have major element compositions that differ significantly from the lavas forming the adjacent volcano, Waianae. The results are somewhat scattered, presumably because of alteration.

Chapter 10

Langmuir   I am surprised that you find a higher total melt production rate for Iceland than for Hawaii. Some previous estimates have been the other way round.

White   The present-day total melt production rate for Iceland, based on the length of the rift, the spreading rate, and the thickness of the crust is 0.24–0.36 km^3 a^{-1}, with the main uncertainty being in the thickness of the crust. If a correction is made for the 10-km thickness of igneous crust that is generated by passive upwelling and decompression melting beneath the adjacent Reykjanes Ridge, then the melt production rate attributable to decompression in the core of the mantle plume is reduced to 0.12–0.24 km^3 a^{-1}. Schubert and Sandwell

(1989) report an average excess melt production rate for Iceland above that of the adjacent ridge of 0.24 km^3 a^{-1} over the past 35 Ma, and my estimate for the period 35–55 Ma represented by the Iceland–Faeroe Ridge is 0.21 km^3 a^{-1}. For Hawaii, seismic cross-sections suggest an average melt production rate of 0.14–0.18 km^3 a^{-1}, in agreement with my estimate of the average for the past 10 Ma of 0.16 km^3 a^{-1} (Fig. 2). So there seems to be little doubt that the total melt production rates for Iceland and present-day Hawaii are similar, but with those for Iceland somewhat higher. Prior to 20 Ma, the production rates for the Hawaiian–Emperor chain were considerably lower still.

I am not aware of any previous estimates that put the total melt production rates for Hawaii higher than for Iceland. For example, Sleep (1990) (*J. Geophys. Res.*, **95**, 6715–36) reports melt production rates of 0.13 km^3 a^{-1} for Hawaii and 0.25 km^3 a^{-1} for Iceland.

Langmuir Davies (1988) (*J. Geophys Res.*, **93**, 10467–80) and Sleep (1990) give ratios of Hawaii to Iceland of about 2 and 6, respectively. Of course, these are for the buoyancy flux, which differs from the melt production. Our knowledge of crustal thickness beneath Hawaii and Iceland is uncertain. For example, the multi-channel seismic data for the Hawaiian swell necessarily run between islands (Watts and Ten Brink (1989). *J. Geophys. Res.*, **94**, 10473–500), where melt production is obviously less. For other islands, there is even less information. Because of these uncertainties, I have used the excess volcanic surface volume as an indicator of relative melt production. The Greenland/Iceland and Iceland/Faeroe rises are about 770 km across with an excess elevation of about 2500 m relative to the surrounding sea floor. Total spreading rate is 20 mm yr^{-1}, yielding a surface volume flux of 0.02 km^3 yr^{-1} assuming a triangular cross-section for the volcanic rise. The volcanic portion of the Hawaiian swell is about 230 km across. To the summit of Mauna Loa there is close to 10 km of elevation above the sea floor. Using an elevation of 7 km, a plate speed of 98 mm yr^{-1}, and a triangular cross-section, the surface volume flux is 0.08 km^3 yr^{-1}.

The critical point is that if the plumes for Hawaii and Iceland were the same, then the melt production should be dramatically different, because of the change in lithospheric thickness. Whether one accepts that melt production is about equal (your estimate), or that Hawaii's is greater (my estimate), I would argue that a single type of plume cannot produce both islands. This argument is supported by the analysis of buoyancy flux, melt production, and basalt chemistry that I presented in my talk.

White The fluxes that you are talking about here are the mantle buoyancy fluxes. These are different from the melt production rates because, as I have discussed, not all the excess buoyancy produces melt. Given identical mantle plumes, the amount of melt that is produced is dependent on the lithosphere thickness. Although the Hawaiian buoyancy flux reported by Sleep (1990) is more than six times higher than that for Iceland, less melt is produced beneath Hawaii because the lithosphere there is thicker than it is beneath Iceland. Mantle plumes come in a spectrum of shapes and sizes and the melt production rates depend both on the parameters of the plume and on the thickness of the overlying lithosphere.

Thompson How have you obtained your estimate of 100 km for the thickness of the lithosphere beneath the Kenyan Rift?

White   This thickness is only approximate, based on estimates of the extension of 10–35 km reported across the Kenyan Rift. In Fig. 4 I have plotted melt production rates from oceanic areas against the square root of the age of the lithosphere. The lithosphere scale along the top of Fig. 4 is an indication of the thickness relevant to the points I have plotted. The Kenyan Rift rate, as the only representative continental result I have shown, is plotted against the lithosphere thickness scale. In fact the precise lithosphere thickness is not important, since the point I wish to make is that melt production rates are very low where plumes lie beneath relatively thick lithosphere: the Kenyan Rift results show this.

Parsons   Your plot for the Hawaiian chain shows the melt production rate decreasing with increasing age. As you remarked, this variation occurs over a time scale of 35 Ma, a typical time scale for convective evolution. But the bathymetry just before the bend in the Hawaiian chain suggests that the melt production rate was higher. How can you account for such rapid changes in production rate?

White   The melt production rate over the period 35–50 Ma decreases at a similar rate to the increase over the period 0–35 Ma (Fig. 2), so it may represent a normal cyclicity in the strength of mantle plumes with a time scale of a few tens of millions of years. The change is no more rapid than is seen elsewhere. But in addition, at least part of the decrease in melt production rate can be attributed to the increasing age difference, and hence increasing lithospheric thickness, of the oceanic crust beneath which the plume lay prior to 35 Ma (Fig. 2).

Osmaston   The Slave dyke swarm propagated for more than a thousand kilometres radially from its source in Arctic Canada. I do not understand how dykes that are intruding laterally can have such constant trends, unless the plate thickness is very much greater than current views admit.

White   The direction of dyke propagation is controlled by the stress field in the lithosphere, and the melt flows away from the elevated areas. Continental breakup above plumes frequently produces radial patterns of dykes, because the uplift associated with the plume imposes a radial stress field and outward flow of melt away from the centre. Often the dyke swarms are in a triple-junction pattern, of which usually only two of the arms evolve to form a new ocean basin, with the third, failed arm being marked by dyke swarms and minor rifting.

Chapter 11

Takahashi   In the Columbia River region the basalts produced at the climax of magma eruption (such as the Grande Ronde stage) are almost totally aphyric. Yet your calculations show that these rocks must have been associated with very large amounts of cumulates. These cumulates are produced at pressures of about 1 GPa. The viscosity of the residual magma increases with the amount of fractionation, and I am concerned whether its viscosity is low enough for it to spread as far as it does at the surface? I also do not understand why such evolved magmas contain so few phenocrysts.

Cox   In the Deccan, aphyric rocks are often interbedded with mildly porphyritic rocks, and in some cases with highly porphyritic rocks. There is no doubt that gabbro fractionation has occurred in the source regions of these basalts by crystal settling. Detailed studies (Cox and Mitchell 1988) show that the difference between aphyric and porphyritic lavas lies in the fact that the latter have Fe-rich, and hence dense, ground-masses. They are the most evolved amongst the basaltic magmas, and their high density enables them to carry crystals to the surface.

Whatever the fluid mechanical explanation may be, it is clear from the field observations that magmas erupted at temperatures of 1130–1180 °C have flowed large horizontal distances.

McKenzie   A basalt with 5 per cent MgO at about 1100 °C travels at about the velocity of an oceanographic vessel, about 3 m s^{-1}, in a dyke whose thickness is 30 m. Therefore melt takes about a week to travel from one end of a large Canadian dyke to the other, and in so doing cools by about 50 °C. Exactly the same calculation applies to a sill 30 m thick. So there is no obvious fluid dynamical problem in accounting for field observations. The Reynolds number in such a dyke is such that the flow will be weakly turbulent or may even be laminar. This result may be important, because the heat transport to the walls is much more efficient when the flow is strongly turbulent. The reason why the dykes can propagate such large distances may be because little heat is lost to the walls. An important calculation that has not been carried out is the influence of strong temperature, and hence viscosity, gradients within the dyke on the Reynolds number at which the flow becomes turbulent. This aspect of the fluid dynamics may control whether or not long-distance transport occurs.

Cox   I think that turbulent transport is quite common. The picritic magmas of the Deccan must start at a temperature of about 1350 °C, but are all erupted at about 1200 °C. They lose heat by turbulent transport to the walls of their conduits as they pass through the crust.

Moorbath   I have never understood how continental flood basalts acquire their frequently quite evolved isotopic and geochemical signature. It is quite different from anything that could be directly derived from a plume.

Cox   Many such basalts have been strongly contaminated by the crust.

Moorbath   But most models of flood basalt generation involve rapid melt production and transport, either to the surface or to the base of the crust. When is there time for crustal contamination to occur?

Cox   In the Karroo, Deccan, and North Atlantic the first, and often violent, eruptions are locally of picritic basalt. These magmas probably move quickly from their source regions to the surface. With time magma chambers develop at depth, and this is clearly one possible site at which crustal contamination can occur. The extent of crustal contamination between different provinces is, however, very variable. For instance, some Deccan basalts are obviously heavily contaminated by older granites. But such contamination disappeared as the eruptions continued, and is not detectable in the Ambenali formation. There is direct evidence for contamination in some of the dykes, which have brought up crustal xenoliths through the volcanic pile. Geochemical observations suggest that such contamination occurs as the magmas are ascending through the crust, rather than while they are in a magma chamber. In the Deccan it is the hottest magmas that undergo most contamination, or the direct reverse of what is expected from the popular assimilation and fractional crystallization (AFC) model. In AFC models contamination is greatest in magmas that have resided longest in the magma chamber. These are also the coldest magmas. In both the Deccan and in Skye, it is the hottest magmas that are the most contaminated. I think that the best explanation of these observations is that the contamination occurs during the ascent of the magmas, and that there is sufficient time for such effects to occur. What is surprising is that contamination does not always occur. In the Karroo, for instance, there is very little.

Moorbath   Though many of the rhyolites and basalts from the Karroo that you showed had little radiogenic Sr, some of the basalts in your figure have $^{87}Sr/^{86}Sr$ of 0.710 or more.

Cox   Yes, those are ones that have been contaminated by the crust. But they are not common in the Karroo.

Harte   Both White and Cox have argued that large amounts of magma pond at the base of the crust. How does this process occur? Magma chambers have been proposed, but what size and shape do they have?

Cox   Wright *et al.* (1989) originally argued that the magma chambers in flood basalt provinces had to be at considerable depths within the crust. They made this proposal because of observations in the Columbia River province, where eruptions of 10^3 km^3 or more are not associated with any surface calderas. The absence of surface subsidence is easy to understand if the magma body was sill-like. A chamber of this shape would also be able to generate large volumes of fractionated magma with very uniform composition. Though I do not really understand the processes involved in such fractionation, it seems to me easier to generate the uniformity in a horizontal chamber than in a vertical one, with large pressure and temperature differences between the top and the bottom.

White   The eruption rates in the largest flows of the Columbia River Basalt Group were 1–3 cubic kilometres per day from each kilometre of the fissures from which the magma was erupted, and individual eruptions lasted for the order of one week. The fissures are up to 150 km long, and produced up to 2000–3000 cubic kilometres of evolved magma in each eruption. These observations require large magma chambers.

Much of the underplated material has a high density and exhibits a high seismic velocity, which allows it to be recognized geophysically. It is solidified magma, probably with a bulk composition similar to that of the original melt. Because the density of such magmas is greater than that of most of the crust, they are trapped close to the Moho and spread sideways as sills or dykes. Only the fractionated portion of the melt has sufficiently low density to erupt as tholeiitic basalt. The melting temperature of the continental crust is only about half that of the basaltic melts, so as the crust is melted by the intruded picritic basalts, it will provide an even more potent density trap. Both geochemical estimates and geophysical measurements of the volume of underplated material on volcanic continental margins suggest that only one-third or less of the melt is extruded through continental crust, with the remaining two-thirds being ponded at the base.

Chapter 12

Gill   I noticed that there was a low-velocity zone on the Pacific coast, between the volcanoes and the trench. How do you think it is produced?

Hasegawa   The structure to which you refer is well resolved by the tomography to an accuracy of perhaps between 0.5 and 1.0 per cent. I do not know why the velocity should be low. Perhaps hot material or water is present there.

White   How much melt do you think is present in the inclined zones, and why does it remain in the inclined zones and not flow rapidly upwards out of them?

Hasegawa   The seismic reflections from the inclined zones allows the velocity contrast to be estimated. The accuracy of this estimate is less than that from the tomography, because it depends on modelling the power spectra of a few seismograms. The velocity contrast is about 1 km s^{-1}, and could be produced by a zone of partially molten material.

Chapter 13

Pearce   You have assumed that the material that comes from the subducted slab is the same in all arcs and carries little cerium. These assumptions are difficult to reconcile with the great variety of material that is being subducted. How can you distinguish the effects of different source region compositions from chromatographic effects in the wedge above the slab? I also do not understand how you can use Ce/Sm ratios to study partial melting when there are cerium anomalies from subduction effects in some of the arcs you have studied.

Hawkesworth   I believe that by chance there are no negative cerium anomalies in most of the data set that we used.

We do not assume that the cerium flux from the slab was small. Instead we assumed that the relative trace element pattern in the fluid was that in equilibrium with mantle peridotite in the arc source, and allowed it to percolate through the wedge, which we assumed started with the composition of a MORB source. The relationship between the trace element contents of the initial fluid in the peridotite and that released from the slab is controlled by the partition coefficients between melt and matrix in the two regions. We did not attempt to model this process, but it can be assumed that all the trace elements in the initial fluid had come from the slab. Those elements with the smallest partition coefficients will produce the most obvious geochemical signatures. We could in principle model the fluid that is in equilibrium with the slab using appropriate partition coefficients and mineral stability fields. The difficulty with such a calculation is that the fluid composition is primarily controlled by these stability fields, which are not yet well known. This is the reason why we did not attempt such a calculation.

Langmuir   Terry Plank, at Lamont–Doherty, has been working on this problem for her doctorate, and has found good correlations between elements like zirconium and crustal thickness, which can be explained by the same model that accounts for the major elements — that the extent of melting beneath an arc is controlled by the thickness of the overlying crust. In your model, the composition of the fluid will be controlled by that of the mantle through which it percolates. Unless the fluid fluxes are large, only those elements with very small partition coefficients can be efficiently transported from the slab to the surface. Terry has also found correlations between the composition of the subducted sediments and that of the associated arc volcanics. So it appears that some of the geochemical signature from the sediments can reach the surface, which means that percolation cannot be the whole story.

I also wondered if you had carried out any calculations on the behaviour of beryllium? In the fluid phase it should have a partition coefficient that is near zero, which means it will not be efficiently transported. In this case, I do not see how your model accounts for the ^{10}Be excesses that are observed.

Hawkesworth We have not done any quantititative calculations on the movement of beryllium. Our aim was first to model the movement of the trace elements, and hence the associated isotopic signature. These calculations show that strontium quickly loses its arc signature as it percolates. In the case of neodymium, the arc signature is anyway very small. The problem with beryllium is that the initial isotopic ratio when the fluid leaves the slab is unknown. It is therefore very difficult to do quantitative calculations.

Bickle Most of the rocks that you have considered are basalts. The fluid flux will produce a melt that is water-saturated. Since the solidus temperature of such melts increases as they move upwards, they will freeze before they reach the surface. Therefore, the mafic melts that are erupted must be produced by processes other than hydrous fluxing; for instance by adiabatic upwelling associated with extension. I wondered how such arguments affected your interpretation?

Hawkesworth The question is whether the observed range in degree of melting can result from plausible amounts of extension. In arcs there is no satisfactory method of estimating the extension. Extensive stretching has occurred in Tonga and the northern Lesser Antilles, but its importance elsewhere is hard to estimate. It is possible that our temperature structure is not correct, and that the amount of water involved in melt generation is less than we believe. A related question is whether the lateral migration of fluid by repeated crystallization and breakdown of amphibole purifies the fluid of ions that do not sit well in the amphibole structure. That, in turn, may change the fluid composition, and hence the fluid/peridotite distribution coefficients so that it is hard to understand how the trace elements can be transported to the region where melt is generated.

Simon Day Your model involves rather large fluid fluxes. I wondered if you had made any estimates of the corresponding melt fluxes, and of how much fluid you need to be released from the slab?

Hawkesworth I have not done so, because the purpose of our model was to evaluate the conditions under which the Davies and Bickle (1991) (*The behaviour and influence of fluids in subduction zones*, pp. 129–30, Royal Society/Cambridge University Press) proposals were possible. It would be much easier to match the observations using their estimates of the flux if the partition coefficients were one or two orders of magnitude less than the ones estimated from available experimental data.

Harte We have been using the ion probe to examine the trace element distributions in rocks that contain co-existing clinopyroxene and amphibole. Our results suggest that the commonly used amphibole partition coefficients for amphibole/melt are considerably too large. Those that we have obtained for the rare earths appear similar to those of clinopyroxene, on the basis that amphibole and clinopyroxene concentrations of REE are quite similar to one another.

Hawkesworth Such a change would bring the patterns together. An important question concerns the behaviour of niobium, and whether the new measurements can account for the observed variations in the niobium–potassium ratio.

Russell Harmon A major difficulty in arcs is the absence of primary magmas. In the central Andes there are no erupted basalts with even 6 per cent MgO. All the rocks have undergone severe fractionation, which is commonly believed to occur near the crust–mantle boundary. Certainly the fractionation must have involved interaction with the crust in both continental and oceanic island arcs. It is also likely that amphibole is one of the minerals that fractionates. Have you attempted to estimate the influence such crustal interaction has on the trace element geochemistry?

Hawkesworth We have not tried to model the influence of such effects on the Ce/Sm ratio. The reason why we used this ratio in the first place is that it is rather insensitive to amphibole fractionation.

Gill You have argued that the melting regime in arcs is similar to that elsewhere, a point that was also made by Takahashi. In his case he could model the arc source by using an even higher degree of melting than is present in the MORB source. Surely such large extents of melting would leave no residual amphibole?

Hawkesworth I agree. What Bickle said earlier was concerned with the initial melts, which are likely to freeze before they are erupted. There must therefore be some other melting process involved. As a geochemist, I would prefer there to be rather little residual amphibole at either stage.

Index